CRAFTS IN THE BARN

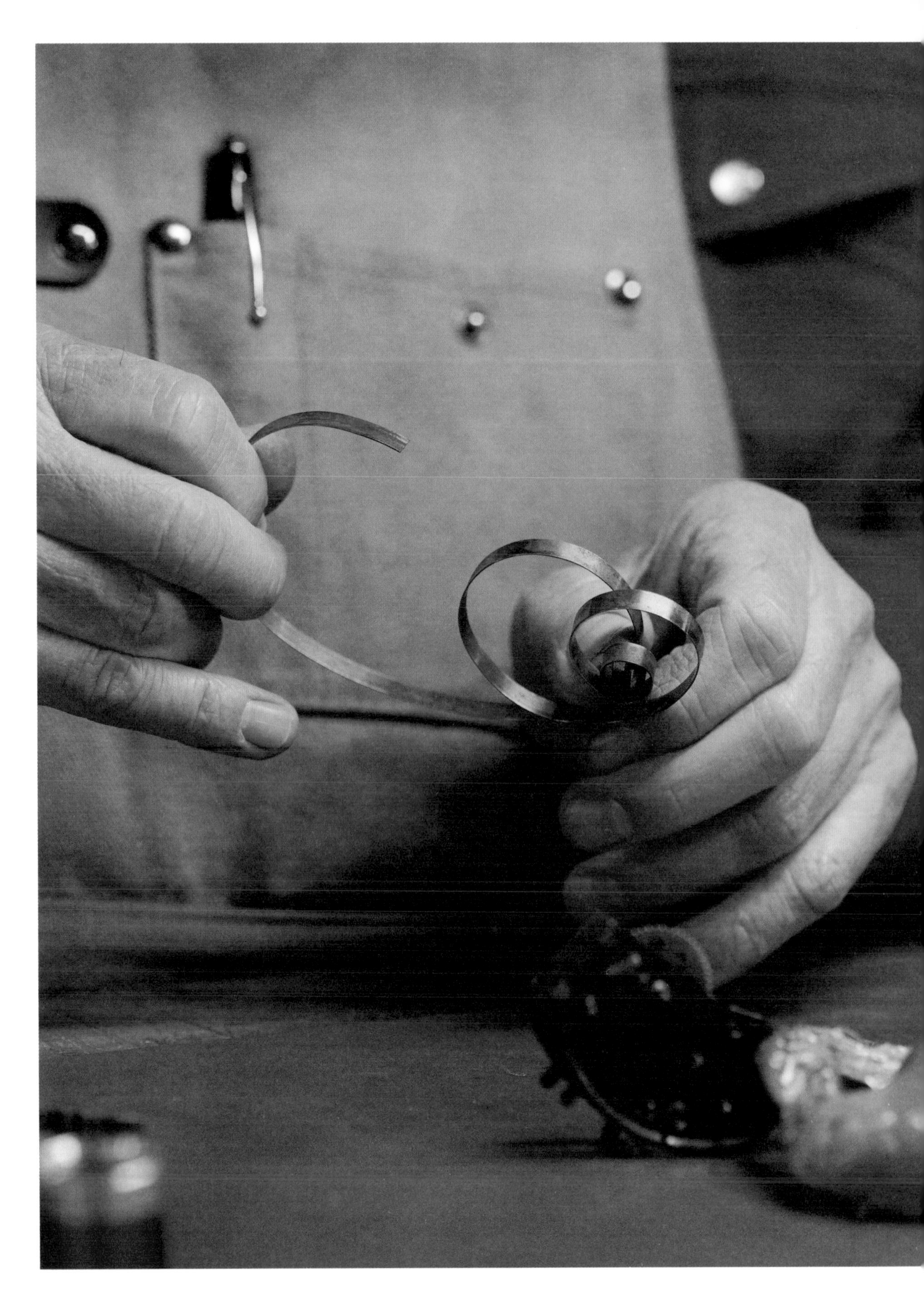

CRAFTS IN THE BARN

Skills, Stories and Heartwarming Restorations

Foreword by JAY BLADES

KYLE BOOKS

An Hachette UK Company
www.hachette.co.uk

First published in Great Britain in 2023 by
Kyle Books, an imprint of Octopus Publishing Group Limited
Carmelite House
50 Victoria Embankment
London EC4Y 0DZ
www.octopusbooks.co.uk

This edition published in 2023

ISBN: 9781914239656

Distributed in the US by Hachette Book Group, 1290 Avenue of the Americas, 4th and 5th Floors, New York, NY 10104

Distributed in Canada by Canadian Manda Group, 664 Annette St., Toronto, Ontario, Canada M6S 2C8

Publisher: Joanna Copestick
Editorial Assistant: Emma Hanson
Design: Rachel Cross
Photography: Sarah Weal
Production: Lucy Carter and Nic Jones

Text by: Jayne Dowle and Elizabeth Wilhide

Printed and bound in Europe

10 9 8 7 6 5 4 3 2 1

Contents

FOREWORD

Jay Blades, MBE

Presenting *The Repair Shop* is a real privilege. I get to work with a whole team of creative people who love craft as much as I do – as well as a whole different team of creatives behind the camera. When family treasures first arrive at the Barn, often in a very sorry state, I have a ringside seat, and I hear at first-hand what these objects mean to their owners, what precious memories they hold, what stories they have to tell. Later, when the items are mended and handed back, I share the same delight and pleasure in seeing them gain a new lease of life. The act of mending is one that repairs people, too. It's an emotional journey, every time.

Lately, I've been a bit busy, I must admit. When I'm not filming in the Barn, I'm involved with a number of organisations that support craft and craftspeople. Back in July 2021 I was asked to become Building Crafts Ambassador for The Prince's Foundation. The crafts programme run by the Foundation trains students in a range of heritage crafts from stonemasonry and thatching to blacksmithing. It's based in Scotland, at Dumfries House, which is where we were invited to go at the behest of the then Prince of Wales.

Last May I was made co-chair of the Heritage Crafts Association, a charity that monitors endangered crafts and promotes their survival. I'm also an ambassador for the Queen Elizabeth Scholarship Trust (QEST), which provides grants for students enrolling in vocational courses. Just recently, I was asked by HRH The Princess Royal to be Vice-President of City & Guilds, the charity of which she is President.

Often I go back to Rycotewood Furniture Centre, in Oxford, where I studied upholstery, to speak to third-year students and visit end-of-year shows. It's one of the venues where we're launching Saturday Clubs during the summer months, where young people can come and dip into some of the crafts to see what they're all about. This project is sponsored by organisations linked with QEST and is free of charge for participants.

What all my endeavours outside *The Repair Shop* have in common is my passion to make the invisible visible. Most crafts had very humble beginnings. For centuries the country was full of worker bees busy earning a living, practising their skills, and handing those skills on to the next generation. But lately many of these heritage crafts have ended up on the 'red list' and a shocking number are in danger of dying out completely. There's also the perception that craft is quite elitist because it has been hidden from view for so long.

I knew about craft when I was growing up, but not in the sense that I understand it now. Once you delve into a subject, it gets quite deep and complex. You have to see beyond the surface. It's a bit like if you decide to take up fishing, and then discover there's a whole world out there, where people who are into fishing make different things. Or if you buy

a table because you really like the look of it, then find out more about the skills that went into making it.

My mission is to bring a new awareness of craft to a wide audience and to make it accessible. I can say to a prospective student: these are your opportunities, there's some money from QEST that you can apply for, you can take a look at some courses that The Prince's Foundation are running, or you can go to university to get the skills you need.

Earlier this year, I was asked to be Chancellor of my old uni, Buckinghamshire New University (BNU), where I studied criminology and philosophy, and where my dyslexia was first identified and supported. I said I would accept on the condition that they reinstated their furniture course and gave me the opportunity to send some people there on one of my scholarships.

My old uni has a campus in High Wycombe, in what was once the centre of the British furniture trade. Ercol, G-Plan, and Parker Knoll all used to be based there, and so was my charity 'Out of the Dark', which was set up to teach young offenders how to restore and repair furniture. It was wonderful to see how the self-esteem of these young people just grew and grew the more they learned, and the more skills they mastered. Imagine being told you aren't going to amount to much because the educational system can't necessarily accommodate you, and then being brought into an environment where you truly flourish. You've made something, and it looks good in your eyes, and then it sells. That's real life, someone paying for your efforts. At school your efforts are marked and given a GCSE grade, or something like that, but a lot of the people we dealt with weren't able to achieve those kind of qualifications because of various learning difficulties. We showed them there was an alternative way of valuing themselves and their own personal abilities.

My particular skills are furniture restoration, especially upholstery, which is how I first got involved with the show. Nowadays, I don't have the time to be so hands-on, but my furniture business, based in Ironbridge, Shropshire, is going from strength to strength, and we've got a great team of people working there.

A new venture is Jay & Co, which I opened in March 2023 in collaboration with Steve Wyatt of Restored Retro, which is based in Poole, Dorset.

Steve is another prime example of an adult who had an addiction problem, recovered, and needed something to focus on. That something turned out to be craft and restoration. I mentored him for a while, and the next thing I knew he had opened a shop and I started to sell some of my stuff there. Then an opportunity came up when the premises next door fell vacant. It was a no-brainer: break a hole through the wall and join the two shops together.

I often get called a 'modern restorer', but my work is more about redesign and reimagining. I like to use pattern and colour in surprising or inventive ways – pattern on the back of a sofa or on an arm of a chair, a pop of colour on a leg. That puts restoration front and centre; it doesn't hide it away. It's like the Japanese art of *kintsugi*, or golden repair, where you repair something like a broken bowl with lacquer or powdered gold to celebrate what's happened to it.

At the same time, you're making second-hand furniture more contemporary for the modern market. Not everyone wants to live with a museum piece in

their house. Keeping it current is one way of giving it some individuality and a new life in our times.

What restoration and redesign is all about is using resources responsibly, repairing and mending, not throwing something on the scrap heap, or sending it to landfill. Craft is a key part of that approach, and it's good to see that a renaissance is starting to get underway. We need people who are going to think differently and sustainably.

'What restoration and redesign is all about is using resources responsibly, repairing and mending, not throwing something on the scrap heap. Craft is a key part of that approach. We need people who are going to think differently and sustainably.'

Jay Blades

Where Crafts Come to Life

Located in the leafy countryside of the South Downs, in West Sussex, *The Repair Shop* barn is where crafts come to life – appropriately enough, since this thatched, timber-framed farm building is itself a stunning survivor from time gone by, now repurposed to a new use at the Weald and Downland Living Museum.

When the barn was first constructed, over three hundred years ago, traditional craft skills were still flourishing and widespread in all parts of the country. Buildings were made by hand, largely using local materials, and so was most of what they contained – from furniture to pottery to cutlery.

With the rise of industrialisation and the arrival of mass production, craft skills gradually began to wane, with some threatening to disappear altogether. Today, with *The Repair Shop* as its showcase, a revival is underway. Week after week the resident experts demonstrate the value of making and mending, as precious family treasures are restored to life, ready to be passed on to the next generation.

Anyone who has ever taken up a craft – whether knitting, crochet, woodworking or pottery – appreciates the deep sense of fulfilment that comes from working with your hands. More than ever, it's becoming a necessary antidote to the fact that these days so many of us conduct a large proportion of our everyday lives remotely, via digital screens. Craft is grounding because it is all about touch, and touch is increasingly a neglected sense.

Whichever craft you choose, you have to be prepared to be patient and put in the time: practice really does make perfect. *The Repair Shop* experts have been on exactly the same journey, starting as outright beginners, becoming apprentices, before ending up as leading professionals in their fields. There's no quick fix.

The skills on display in the Barn have not been acquired overnight, but are the product of hours and hours of trial and error, of figuring out what works, of experimenting and making the occasional misstep and correcting it. And the learning is lifelong. Many of the objects that come into the workshop present challenges that the team have never encountered before, which is when two – or three – heads become better than one, and lateral thinking, as much as an individual's specific technical knowledge, is the order of the day.

Craft is a state of mind. At its best and most fluid, the connection between hand and eye promotes an almost meditative feeling, a heightened sense of focused concentration that is often called 'flow'. For woodworker Will Kirk it's sanding that brings that element of simple enjoyment, which is also how Jay Blades gets in 'the zone'. Experienced craftspeople know better than to rush such repetitive tasks, and instead are prepared to devote as much time to them as is necessary. While the most dramatic stage is often the final transformation – for example, the coat of paint, varnish or polish that brings everything into

sharp focus – its impact is dependent on the hours of painstaking preparation that went on beforehand.

If there is satisfaction to be gained in the craft process itself, both making something new and repairing something old and time-worn are equally rewarding, not least because the results are tangible. Giving a battered table a new lease of life extends its practical use and prevents what otherwise might have been discarded from ending up on the scrap heap. Cleaning layers of disfiguring varnish off an old painting restores its beauty and honours the artist's original vision.

And it goes without saying that mending not only represents an economical use of resources, both financial and material, it's good for the planet, too. When you fix an object rather than throw it out or rush to buy a new one, you reduce waste and consume less, strategies that are becoming ever more critical at this time of global climate change.

MATERIAL QUALITY

To a great extent, the inherent characteristics of a particular material dictate how it can be worked. This goes far beyond superficial appearance and embraces qualities such as density, pliability, resistance, how it performs under various conditions, and what adversely affects it. Drop a ceramic bowl onto the floor and it's likely to chip, crack or shatter, which is not the case with a leather bag. Prolonged exposure to the elements will cause metal to rust; damp isn't great for wood, either, while other natural enemies include dry rot and woodworm. Learning a craft is not simply about mastering a range of techniques; it's also necessary to get a feel for materials on a fundamental level.

Using quality materials is important to ensure that objects are repaired to a high standard to be treasured for years to come.

Just as a potter will come to appreciate the malleability of clay and how the centrifugal force of the wheel will help shape the final form, the metalworker will gain a sense of the degree to which metal can be bent, hammered or heated to achieve the desired result without breaking, cracking or snapping in two. In every case, it's about letting yourself be guided by the material.

TOOLS

On the other side of the craft equation are tools. Different crafts have their own specialist tools designed to do particular jobs, such as robust, curved leatherworking needles, or the smooth bone folders used in bookbinding. Yet many are surprisingly multipurpose and can be pressed into service to fulfil a variety of functions, sometimes even interchangeably across a number of different disciplines. Few craftspeople are without at least several pairs of pliers.

Screwdrivers, hammers and punches are all tools that embody a kind of working intelligence and become thought in progress. The best feel right and balanced in the hand, although this is often a matter of personal preference – what feels right to one person won't necessarily suit another who has a different grip or degree of manual strength.

Good-quality tools of all descriptions are built to last, which is why many of the experts in the Barn go to great lengths to seek out tools from second-hand sources, such as junk shops and car boot sales. It's also why they treasure those that they first acquired decades ago, which are so familiar they become like extensions of their own hands. They are careful to maintain their tools, which may entail sharpening, oiling or cleaning after use. This is not merely to ensure the tools last longer and function at maximum efficiency, it's also a question of preventing damage. You always have to be alert and careful when you're working with sharp tools, but blunt tools can be more even hazardous – you will be tempted to apply more pressure for the same effect, which will increase the risk of the tool slipping and harming yourself or the material – or both.

Many of the *Repair Shop* experts have their favourite tools; in some cases these have been handed down by members of their own family who were engaged in the same craft. Clockmaker Steve Fletcher and leatherworker Suzie Fletcher, who are siblings, both use tools that once belonged to their grandfather. In Suzie's case, this is a pair of ordinary pliers, while Steve treasures a simple rachet screwdriver. Neither of these tools is particularly special, or is worth much; it's the memories they enshrine that make them irreplaceable.

The experts often use the same tools for many years and they become like extensions of their own hands.

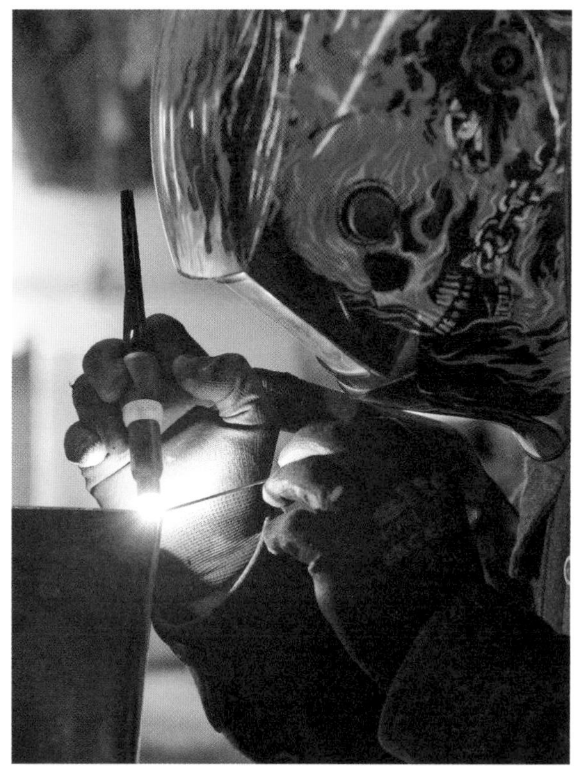

Preserving Heritage Crafts

In 2017 the Heritage Crafts Association published a report listing crafts that are endangered in the UK, including those that didn't originate in the country. The *Red List of Endangered Crafts* was the first of its kind and has since been regularly updated. It makes fascinating, if alarming, reading.

Crafts are grouped into four categories: extinct in the UK, critically endangered, endangered and currently viable. A heritage craft is defined as: 'a practice which employs manual dexterity and skill and an understanding of traditional materials, design and techniques, and which has been practised for more than two generations.' It is deemed currently viable if there are enough people still practising it to pass on their skills to the next generation, along with enough would-be apprentices queuing up to learn it.

Years ago, when Jay Blades first set up a charity teaching young offenders how to restore and repair old furniture, he had no craft skills himself to draw on. So he did the sensible thing: he contacted Age Concern (now Age UK), the Women's Institute and Neighbourhood Watch and asked for volunteers to come and teach their skills. The charity was based in High Wycombe, which used to be the centre of the British furniture-making industry, and Jay was flooded with offers of help. Their oldest teacher was a 92-year-old, who passed on the crafts of caning and making rush seating.

Crafts become endangered and threaten to disappear altogether for various reasons. One of the most obvious is demand, which is closely related to market value. Before the arrival of mechanisation, horsepower was the main engine of transport and agricultural work, which meant the related crafts of saddlery, harness-making and blacksmithing thrived in every village and parish. The fact that none of these crafts are in danger today, despite our reliance on the car as a means of getting from place to place, is because riding horses for pleasure or sport continue to be popular activities.

Other crafts die out because artificial or manufactured products are perceived as better, or are cheaper and easier to produce in quantity. This may account for the reason why making lacrosse sticks and cricket balls by hand are crafts that now fall into the extinct category. What is more worrying is that those on the endangered list include brush-making, marbling, clockmaking, watchmaking, cutlery making, glove-making, hat-making and shoemaking, among far too many others.

Why does this matter? Mass production and marketing means you can buy the same products in Cornwall as in Carmarthenshire, in Norfolk as in Northumberland. Often this standardisation comes at the price of cultural loss. The making of Sussex trugs and Devonshire stave baskets are endangered crafts, and their departure – and the departure of other crafts like them – would mean that regions up and down the country are a little less distinctive and a little bit more like everywhere else. This is because

when heritage crafts die, so do local traditions and diversity.

Perhaps more importantly, the loss of a craft represents a direct loss of knowledge; not only in how to make something by hand, but how to repair it, too. This is where *The Repair Shop* comes in. If you know how to make something, you know how to restore it so that it can be used and enjoyed for years to come.

All of the above are good arguments for learning a craft yourself. This book will introduce the basic skills involved in a range of disciplines and features a number of easy repairs that beginners can attempt. But if getting involved in learning some of these techniques isn't your thing or feels too much, at the very least, we hope you will see your old pieces in a different light. If we relish objects or products that are handmade from natural materials, we ought to be prepared to pay a little more for them to keep these skills in demand and craftwork of all kinds flourishing.

Metalwork expert Dom Chinea and ceramics conservator Kirsten Ramsay in The Repair Shop Barn. Here Kirsten is working on the Tower of London Poppy.

ART RESTORATION

ART RESTORATION

Lucia Scalisi

There's a fine line between restoration and conservation, and this is particularly true in the case of artwork. Restoration implies returning something to its former glory, while conservation aims to extend the life of an item and preserve it for future generations.

A dilemma arises when the object in question is to be used in some way, not just put on a shelf or a wall to be displayed, or tucked away in a drawer as a keepsake. For example, when a chair with a broken leg needs repairing so that it can be sat on again. Making an object serviceable, without eradicating all the evidence of its previous history, requires many fine judgements.

Art has no such practical uses: its purpose is to be admired and appreciated. Yet there is still a tightrope to be walked between doing too little and doing too much. Lucia Scalisi, *The Repair Shop*'s resident art conservator, is more than capable of restoring an eighteenth-century painting so it looks like it was painted yesterday. But she wouldn't, for reasons that are both ethical and aesthetic. To do so would be to compromise the artwork's integrity, and put an inappropriately fresh face on a painting that has seen over 200 years' worth of history. Her principal intentions, instead, are to prevent further deterioration, to mend tears and rips, and to retouch principally those areas where it is necessary in order to make sense of the image. Most especially, it is to clean away layers of disfiguring dirt or varnish that obscure colour, modelling and detail so that the art can be revealed in all its original subtlety.

The distinction between art restoration and art conservation is further underscored by the different types of training involved. Art restoration, certainly in the past, was chiefly apprentice-led, while art conservation is a formal postgraduate course offered by various universities. As a discipline, the latter originated with courses begun in the 1960s under the aegis of institutions such as The National Gallery. Along with an appreciation for art, a requirement of such work is a certain aptitude for science, particularly chemistry.

Lucia can usually tell roughly how old a painting is pretty much straight away. In this she's aided by her in-depth knowledge of which pigments were used during which period, and her familiarity with different historical styles of painting. She likens this to the ability of a devotee of fashion being able to tell at a glance whether a dress dates from the 1930s or 1950s based on the cut and fabric pattern. For example, white paint used to be made from lead, and when lead was identified as being toxic, in the 1920s titanium white replaced as a safe alternative.

When retouching paintings, Lucia works with a basic palette of about sixteen colours, at least half of which are earth pigments. Earth pigments, such as burnt and raw umber, burnt and raw sienna, black and ochre have been used since prehistoric times and are very stable, which is why cave paintings have lasted for thousands and thousands of years.

In the past, these pigments would have mixed with binders such as blood and wax. The modern synthetic pigments Lucia uses range from phthalocyanine, a bright, rich blue, to alizarin crimson, which mimics the red produced by the madder plant.

Older pigments often had problems with stability and toxicity; Prussian blue, for example, an eighteenth-century pigment, looks very different under natural light than it does under artificial light, a phenomenon known as 'metamerism'. Mercury, which used to be the source of vermilion, is toxic.

You need surprisingly small amounts of pigment for conservation work. Lucia sources hers from Kremer, a leading manufacturer based in Germany, in modest quantities of 50g (1¾oz) to 100g (3½oz) at a time. In the case of ivory black, 100g (3½oz) will last a lifetime.

All paint consists of a pigment and a medium to bind it to a surface. Lucia mixes her finely ground pigments with a modern transparent synthetic resin. Like all art conservators, she doesn't use oil, because this takes up to a year to dry, and the colour shifts in tone during that period, which is problematic if you are trying to match areas of original paint in your retouching work.

When a painting or artwork first comes into the Barn, Lucia will inspect it closely, looking for areas of flaking paint and damage to the canvas or support. She'll also use a wetted cotton swab to test a small area for surface dirt and to determine what type of varnish, if any, has been applied. Signatures and dates on the painting itself, along with labels and dates on the back of the canvas can provide important clues as to provenance.

If the canvas is torn or damaged and if there are areas of flaking paint, these repairs are made first. Lucia will use her lining iron to back a torn or frail support with additional canvas, while the heated spatula comes into its own for flake-laying. She places acid-free tissue over the flaking area, feeds adhesive through it, and gently smooths down the flakes with her fingertip, before using the spatula to dry the adhesive.

Cleaning, which is the next stage, can be a complex and monumental task, and it is easy to see why it isn't something that amateurs should ever attempt themselves. To clean off surface dirt, Lucia uses distilled water, adding a drop of ammonia to break the surface tension. This leaves no residue behind, unlike a surfactant. She will use this to carefully go over the entire painting with cotton swabs.

Most of the paintings that arrive in the Barn have been varnished with natural resin derived from a plant source. This type of varnish discolours in less than a decade, obscuring the original colours and modelling, but fortunately it is soluble and is relatively straightforward to remove.

However, complications can arise when a new layer of varnish has been applied over an old varnished layer that has become embedded with dirt over the years. Another factor to consider is that certain pigments, particularly glazes, tend to be more solvent sensitive.

Polyurethane varnish, which was very popular in the 1950s, is a real nightmare to remove. Polyurethane 'crosslinks' rapidly, which is when molecular bonds break down with age. This means that only a few years after its application, the surface will appear foggy and grey.

Lucia can tell roughly how old a painting is straight away with her in-depth knowledge of which pigments were used in which time periods. .

Yet another variable is when a picture has been overpainted. With older paintings, especially portraits, it's not uncommon for details to be painted out or altered over time. Uncovering these layers to reveal what lies beneath is all part of the detective work.

A key element of art conservation is reversibility. Once the painting has been cleaned of surface dirt and old varnish, Lucia brushes on a new layer of varnish. This isolating layer protects the original work and means that subsequent retouching and filling can be undone in the future, if necessary or desired. Modern synthetic varnishes, such as those that Lucia uses, have great ageing properties and transparency.

Traditionally, filling was done with a putty-like mixture of gesso and animal glue. Lucia uses water-soluble filler, mixed with a dentist's tool, to provide a base for retouching. Afterwards she will brush or dot on an opaque base coat to the fillings. Then and only then can the retouching begin.

Lucia never retouches over the original paint. Her aim, instead, is to fill in enough missing areas to enable the image to be properly read and appreciated again. It's exacting work and demands a keen eye for matching colours. Using only a tiny amount of paint at a time and the very finest sable brushes (0 or 00 Winsor & Newton series 7), Lucia carefully applies the relevant colour, taking pains to copy the original style of brushwork so the retouched areas blend in seamlessly. The very last stage of this meticulous process is a final coat of varnish, this time sprayed on, rather than applied with a brush.

MY FAVOURITE TOOL

'I've worked with my heated spatula all my professional life. I bought it after I finished my conservation degree and first started working at the Victoria & Albert Museum. Specially made for conservators, the spatula is attached to a control unit, which enables the temperature to be set at different levels. You can also plug in an iron to the back for lining paintings. I mainly use it for flake-laying and consolidation. The temperature will be dictated by the type of adhesive I'm using at the time. A heat-seal adhesive, for example, will only flow when the temperature is 65°–70°C (149–158°F).

From time to time I'll send the unit back to the manufacturers for recalibration to ensure the thermostatic control is absolutely accurate. Appropriately enough, Willards, the company that makes the tool, is based just a few miles from the Barn.'

French Portrait

Famous ancestors brought back to life

Father and daughter Jean-Robert De Bisschop and Natasha Walker-De Bisschop arrive in the Barn with an intriguing family portrait in need of some serious TLC from painting conservator Lucia Scalisi.

This oil on canvas portrait of Jean-Robert's father, Alain, when he was a little boy connects this British-French family with their famous ancestors, the de Beauvoirs. It was painted by Alain's cousin, Hélène, an artist who passed away in 2001 at the age of 91. Well-known in her own right, she was also the younger sister of famous French philosopher and pioneering feminist, Simone de Beauvoir, and a strong advocate of women's rights herself.

The painting is not signed or dated, but the family believe it was painted in approximately 1939, when Alain was two or three years old, and when Hélène escaped to Portugal after the Nazis invaded France during the Second World War.

Hélène painted various family members, but she never signed family pieces. Although the painting is not signed, it is very valuable to her cousin's descendants.

Natasha, who was only eight when her grandfather died, says that to her, the painting represents her family's history. She was too young to ask questions of Alain – her 'Poppy' – but has learned many stories about his fascinating past.

The painting had hung on the wall behind a door to the sitting room in La Grillère, the family farmhouse near Limoges, for as long as anyone in the family can remember. Fondly referred to by the family as Poppy's House, this remote homestead had been enjoyed as a holiday home and the place for huge family gatherings since Natasha was a child.

Jean-Robert lives in France, in Toulouse, and Natasha moved to Nottingham in 2018 to study. Alain passed away in 2008, and the house was eventually sold in 2022.

Alain was very much loved and missed. His family were so distraught at his death that, they couldn't bring themselves to go to the farmhouse to retrieve the painting, knowing that he would no longer be there. Eventually, though, Jean-Robert

The treasured French Portrait reminds the family of a much-loved father and grandfather.

collected it, and in November 2022, Natasha saw the painting for the first time since her childhood, when her dad brought it over to the UK for her. Natasha was saddened to see the extent of the damage and admitted that the entire family felt guilty for not retrieving the painting sooner and allowing it to deteriorate so badly.

The painting has suffered a lot of water damage, and the paintwork is faded, with the paint so thin in places that it has disintegrated. There are also smaller holes, which may have been caused by woodworm in the past – although there is no active woodworm present.

Lucia is intrigued by this painting. It's a real rarity to have in the Barn, a piece of social history that, had it not been kept safe within the family for so long, would have been lost or forgotten about. Her first observation is that it has suffered huge losses of paint, so much so that there are areas where the canvas beneath is completely exposed. Also, there are a lot of horizontal weft fibres missing from the canvas material itself.

Lucia's first priority is to replace the lost fibres. She finds there are some excess threads at the side. So Lucia delicately unpicks the weft from what is spare, and weaves these strands into the damaged canvas to fill the gaps.

After using a filler to create a flat, smooth surface on the canvas, she carefully paints over the missing bits of the portrait. The end result is illuminating; Jean-Robert is amazed when he returned to the barn; saying it was as if his father had come back to life.

Lucia replaces lost fibres on the canvas and applies a final coat of varnish to the treasured French Portrait.

Trimurti Portrait

A lenticular is brought back to life

A fascinating three-dimensional portrait from India holds great religious significance, but just as importantly for Raj Joshi, it keeps alive the memory of his creative grandad and much-missed mum, who both passed away in India.

The portrait is a lenticular, a multi-image Hindu depiction of God (Brahman), represented as many gods, in a style of portrait known as a 'Trimurti'. This is a Sanskrit word which translates in English to 'Three Forms'. In Hinduism, 'Trimurti' is the triad of three gods, who create, preserve and destroy the world.

The 3D design means that when you look at the portrait from different angles you can see different figures. At first, it looks as if there are only three gods, but the fascinating thing about this lenticular is that there are actually five present.

Raj, from London, tells us that when you regard the portrait straight on, you'll see the original Trimurti. From left to right are Brahma (the creator), Vishnu (the preserver) and Shiva (the destroyer). However, if you look at the portrait from the left, you'll also see Saraswati, the female counterpart of Brahma, and the goddess of music and learning. Look at the portrait from the right and Lakshmi, the female counterpart of Vishnu, the goddess of wealth, comes into view.

Raj thinks the portrait is stunning, quite inventive and creative in terms of the depiction of Hindu gods. He doesn't know for sure that his grandad painted it, but the family story goes that somehow he must have cut an existing portrait into strips, then concertinaed these slats together to create the 3D effect. He says that here is a similar Trimurti portrait in the Haripur Mandir ('Mandir' is Hindi for Hindu Temple, and Haripur is a village in Maharashtra, India), that can still be seen today. This, he believes, inspired his grandad, Sri Ganesh Anant Deodhar, a Pujari (Hindu priest) who was born in India in around 1890, to make something similar.

There are no existing photos of Raj's grandad, because there were no cameras in the Haripur village at the time when he was alive. So this portrait is the only visual memory the family have of their beloved ancestor.

When Ganesh passed away in the 1930s, the portrait was passed on to Raj's mother, Kusum

This fascinating three-dimensional Trimurti Portrait holds special memories of a grandad and much-missed mum.

Viswas Joshi (née Ganesh Deodhar), who was born in Haripur in 1931. Although she eventually moved to the UK and the portrait belonged to her, it stayed with family in India.

However, forty years later, after retiring from her job in government administration, she went back to India with Raj's father, retired accounts clerk Vishwas Ramchandra Joshi, and was reunited with the portrait. Kusum was delighted to have this heirloom back in her possession after so long, as it kept her close to her father.

Kusum died in 2019, but before she passed away, she bequeathed the portrait to Raj. He took it back home with him to the UK, where it has been a constant reminder of his mum ever since. However, as it is now nearly a hundred years old, it's difficult to make out the images themselves. Raj is worried about the portrait deteriorating further and wants to see it as it was when it was first made.

This portrait holds great significance to him because of its link to his mum and grandad. Raj says that for him, to see it how it was intended to be would add to its beauty. If it was repaired, he would put it up in his house, somewhere he could see it each day and think of his mum. And when he passes, he will give it to his son, carrying it down three generations.

Painting conservator Lucia Scalisi is also very knowledgeable about the piece's origins. She has spent time in India, and stresses that such a religious item – illustrating the creation, maintenance and destruction of the world – carries a lot of weight in the Hindu community.

FRAMING

'Framing not only protects a painting or artwork and allows it to be handled safely, it also helps to set it off from its surroundings. There's nothing wrong with a mass-market frame for something like a poster, but for anything of greater value – sentimental, historical or otherwise – it's best to get it professionally framed. A framer can help you select not only the frame itself but also advise whether or not the artwork would benefit from a mount. I work with a framer called Derek Tanous whose family has been in the trade for three generations.

While simple modern frames tend to suit contemporary artwork and more ornate, carved and gilded period styles go well with older paintings, it's really a matter of taste. Sometimes a contrast can be very effective.'

It's a fascinating one-off piece, she says, a very interesting 3D optical illusion created by a simple but clever technique that actually involves two identical images being sliced up and concertinaed back together.

However, it now needs some serious attention. The metal plate on the back is coming off, which means that inside the glass that protects the portrait has become full of dust and obscured.

The first thing to do is take off the metal plate, which Lucia de-rusts, gently removing the accumulated debris. The concertinaed slats are all double-sided, with each side bearing a print of different images, and have probably been glued together with animal glue. Over time this has crystallised, darkened and crumbled off, and the slats have become distorted with age. Lucia needs to go through every single one and reglue them back together so they work effectively again.

She reattaches the slats using an EVA (ethylene-vinyl acetate) conservation-grade adhesive. She also re-touches with watercolour paint any missing or faded pigment elements. Finally, Lucia also re-touches the dented and scuffed black frame, giving Raj's treasured family memorial the flourish it deserves.

Lucia carefully retouches any missing areas of paint and reattaches the slats to the painting, as well as retouching the frame.

'I find sitting and retouching very meditative. To help me focus and get in the zone I'll play the same piece of music on a loop all day, either Indian bamboo flute music or a Bach cello concerto. I've done this for years!'

Lucia Scalisi

Lucia retouches the frame for the Trimurti Portrait to restore this special historical painting to its former glory.

COLLABORATION

Suffragette Desk

A small desk with a big history

This desk once belonged to a trailblazing suffragette who scandalised Edwardian society by refusing to 'obey' her husband, and it is the proudest possession of Sophia Harwood Murday from Dumbarton, who arrives at the Barn with her aunt, Susie Grandfield, from Hampshire.

This delicately decorated, roll-top, bureau-style satinwood desk, believed to have been made around 1900, was originally owned by Una Duval (née Stratford Dugdale), who was Susie's grandmother and Sophia's great-grandmother.

The desk lid and drawers are decorated with a painted 'Rococo' style floral motif and a pastoral scene of a family enjoying a rural picnic. However, behind this genteel scene is a tough battle. Both women love the desk because it represents the effort and achievement of Una and her suffragette sisters in arms. There is a lot of history wrapped up in the desk and it's not just family history, says Susie, it's also part of our collective history: votes for women, the suffrage movement and political campaigning.

Una, who was born in 1879 really was a remarkable woman, Sophia says. As a leading member of the suffragette movement at the beginning of the twentieth century, alongside her husband, Victor, who was also a passionate campaigner for women's rights, she would organise and attend marches and protests to speak in favour of what was known as 'universal suffrage' – the right of all adults to vote and, in their case, women.

Una organised two speaking tours for Mrs Emmeline Pankhurst, the leader of the suffragette movement, probably while sitting at this very desk and writing letters to fellow campaigners, reading her correspondence and writing her diaries.

Sophia and Susie treasure the artefacts that their courageous ancestor left them, which bring home the struggles she faced. It's really quite inspiring how determined they were, Sophia says, explaining that some of the activities they were involved in were dangerous.

The government of the day were aggressive in trying to quash suffragette meetings and protests, passing a number of laws that made it more difficult for the campaigners to protest. Una herself was imprisoned for a month after one such protest.

In an interview, Una relates: 'There was much in prison to unnerve one, but I vowed when I went in, I would never shed a tear. And when I came out, I realised to the full, ever so much more, how important it was for women to have political power.'

The bravery of the suffragettes is still clear; in one of the drawers is a sash that Una wore to a protest in 1910. It has brown stains that Sophie explains are blood; Una and her suffragette sisters were beaten by the police. Everyone in the Barn is shocked by this.

Una left the desk to Susie in her will, who passed it to Sophia, her eldest niece, in 2020. Sophia, in turn, will leave it to her own eldest niece, Elsa.

Susie actually remembers her grandmother sitting writing at this desk, and she demonstrates how the section lifts up and out to reveal the writing surface.

Left, the 'Rococo' style floral motif and pastoral scene of a family enjoying a rural picnic on the lid of the Suffragette desk.

When Una and Victor decided to marry, Victor felt very strongly that Una should not use the word 'obey' in their wedding vows, as he believed that women should not have to obey men.
Una applied to the church to see if she could get a special dispensation to not say the word. On the morning of her wedding she received a call from the vicar. He told her he had spoken to the Archbishop of Canterbury, no less, who informed him that Una had to say 'obey'. So, as Susie recounts, she had to find a way of doing it without uttering the actual word.

This was achieved 'by way of a shuffle and a cough', and caused much scandal, with newspapers reporting on this society wedding where the wife refused to obey.

Una continued as a social justice campaigner for the rest of her life. Susie and Sophia have some of her letters from the 1950s which she wrote while campaigning on behalf of black women in South Africa.

Over the years, the delicate paintwork on the front of this important desk has rubbed off. Susie and Sophia would be delighted if this could be restored. Structurally, the desk is fairly sound, but there is a vertical split on the left side and some of the beautiful decorative knobs inside the drawers are missing.

Now this desk, which saw many a protest letter penned in the name of the fight for votes for women, will begin its long journey back to restoration. Lucia is tackling the damaged paintwork, and Will has the woodwork in hand (for more on this, see Will's fix on page 50).

Art conservator Lucia says the desk is a beautiful, elegant piece and she can't wait to have a closer look at it. She is impressed with the skill involved in creating the paintings on the surface and feels that whoever painted them was very talented. However, the desk is quite badly damaged.

Before beginning work on the restoration, Lucia researches how the intricate details of both the floral motifs and the pastoral scene have been depicted by the artist. Using various references from the Rococo period, she settles on a depiction of a woman extending and placing a floral wreath on the head of a child.

Handling and dusting the surface over the years has contributed to cracking and losses of paint. There have also been various coats of varnish added, which will have dulled the once-vivid sheen of the Rococo flourishes.

Lucia uses dry pigments in synthetic resin to retouch the paint losses, employing a pointillist technique to dot on the colour and re-create the missing sections of paint. This means she has to painstakingly dot in the paint to bring the design

PROTECTING PAINTINGS AND OTHER ARTWORK

'Heat, light, dust and dirt will discolour artwork and break down the paint layers. For this reason, it's best not to hang paintings over working fireplaces, or in direct sunlight, or over a radiator, where they will be adversely affected by convection currents.

Framing is also an important means of protection, especially when combined with glass. Even so, dust and dirt can creep in if the backing is torn or damaged in some way.

Above all, if you're tempted to have a go at cleaning a painting yourself: don't. Without a precise knowledge of the chemical composition of the paint, support and materials, you're guaranteed tears before bedtime.'

together. It is hard to do justice to this very special desk, she says, admitting that she has really got her work cut out.

Starting on the most obvious areas, such as the chocolate-brown-coloured wooded background, Lucia focuses on turning back the whole area into as pretty a picture as it was intended to be.

When the delicate process of bringing all the images back to life is complete, she adds a light spray of modern synthetic resin varnish to seal her work.

Lucia carefully restores the painted image on the lid of the Suffragette Desk, which is a huge inspiration to Sophia and her family.

WOODWORK

WOODWORK
Will Kirk

***The Repair Shop*'s resident woodworker since the very first series, Will Kirk has had the opportunity over the years to restore and repair a wide range of family treasures, from handmade model boats to clock cases, along with many pieces of battered furniture so dilapidated that they otherwise would have been destined for the scrapheap.**

As always, there are fine judgements to be made. If an object has a functional use, it's important to repair it so that it's returned to working order – so you can sit on a chair and it won't wobble, or so that a table isn't rickety and threatening to collapse. But because wood acquires a beautiful patina and depth of character with age, Will is careful to respect the history of all pieces and not remove every trace of the passing of time, including the odd scratch, nick or gouge.

Wood is an incredibly diverse material, varying in colour, grain, texture, density, availability and cost. Excluding manufactured woods like plywood and chipboard, there are broadly two categories: softwoods and hardwoods. Softwoods are derived from coniferous trees such as pine and spruce, which tend to be fast-growing, sappy and knotty. These economical woods have a vast range of applications in construction and furniture-making. More expensive and denser are hardwoods, which come from slower-growing trees, some of which are deciduous, some evergreen, and are found in both temperate and tropical regions. Ash, birch, beech, oak, maple, teak, mahogany and walnut are all types of hardwood that have frequently been used in furniture-making. Since many types of hardwoods are expensive, particularly those that are most exotic, they are often used as decorative veneers over a cheaper material.

As a practice, woodworking can generally be divided into carpentry and cabinetmaking, although the relevant skills and techniques do overlap. While carpentry ranges from making structural foundations and frameworks, such as stud walls and staircases, to simple projects such as putting up shelves, cabinetmaking is more refined: here's where a set of shelves becomes a bookcase, for example. Both types of woodworking require a familiarity with a range of both hand and machine tools to carry out basic tasks such as drilling, sawing, hammering, smoothing with a plane, and fixing using nails, screws and glue. More advanced are the skills required to create joints, from simple mitres that you might find at the corners of a picture frame, to mortice-and-tenon, and, the height of finesse, the dovetail. Similarly, all wood needs a protective finish if it's going to be exposed. This may be a coat of paint or vanish, or in the case of hardwoods, French polishing, which is an art in itself.

While a number of traditional crafts associated with working in wood, such as making cricket bats,

oars, spars and masts, are on the endangered list, what is equally worrying is that many types of wood are, too. It's not simply that various species are at risk of extinction because of over-harvesting for lumber, but that other flora and fauna in the habitats where the trees grow are similarly impacted, along with the way of life of indigenous communities. Restricted or endangered woods include many exotic species: afrormosia, Brazilwood, cedar of Lebanon, ebony, iroko, wenge, satinwood, some varieties of mahogany and most rosewoods, to name but a few. Despite attempts to ban, or at least regulate, the trade in these woods, illegal logging still goes on in many areas of the world, putting local ecosystems at risk. All wood used today should either be reclaimed from pieces salvaged second-hand – a strategy that has enabled Will to accumulate an invaluable collection of veneers for his restoration work – or FSC-certified. The FSC (Forestry Stewardship Council) label verifies that the wood has come from sustainable sources.

Will hadn't even realised that you could do a course in antique restoration until his mother spotted an ad in a local paper and pointed it out to him. Nowadays, he tries his best to bring the possibilities for working in craft to the attention of other young people, who, like him, don't fit easily into an academic mould. Along with other *Repair Shop* experts, he's an ambassador for the Heritage Crafts Association, and a supporter of various initiatives to raise funding for apprenticeships, some of which are run by The Queen Elizabeth Scholarship Trust (QEST). With woodworking no longer offered as practical subject in most schools, and college courses increasingly under threat due to lack of funding (including the one at his old uni), Will is keen to make crafts as accessible as possible.

Dexterity and patience are obvious qualities required by anyone who earns their living working with their hands. For Will, the ability to problem-solve is also crucial, as is the case with the other experts in the Barn when they are presented with a type of object they have never encountered before. The beginning of a new project will often find him staring into space; despite appearances, Will is busy working, consulting the library in his mind to come up with the ideal solution to the puzzle he's confronted with. At the same time, he'll be evaluating the wood, its grain, how hard it is, and whether or not it is veneered.

One of Will's special skills is French polishing. This needs a perfectly prepared surface, which in the case of a well-worn piece of furniture means cleaning and sanding. To remove surface grime, he'll use a soft cloth and a mild solution of turpentine and white spirit. Next he'll remove old varnish using methylated spirits and either a soft cloth or very fine steel wool. Then it's a question of sanding, followed by more sanding, working from the coarsest grades to the finest.

French polishing, which gives a rich, glossy patination with great depth, is built up by using a pad to apply successive thin layers of shellac mixed with white spirit. Pigments are mixed into the polish to match the colour of the existing wood grain.

Will is especially careful when it comes to measuring and cutting. The old adage 'measure twice, cut once' translates to 'measure a billion times' when he has one shot at using a piece of veneer that is the perfect match for a repair. After the veneer is cut to the right shape, it is glued into place and secured with a clamp until it dries.

Carving is a form of woodwork that Will especially enjoys. You need a sculptor's eye to visualise the form within the block of material, but the satisfaction of completing a piece is hugely rewarding.

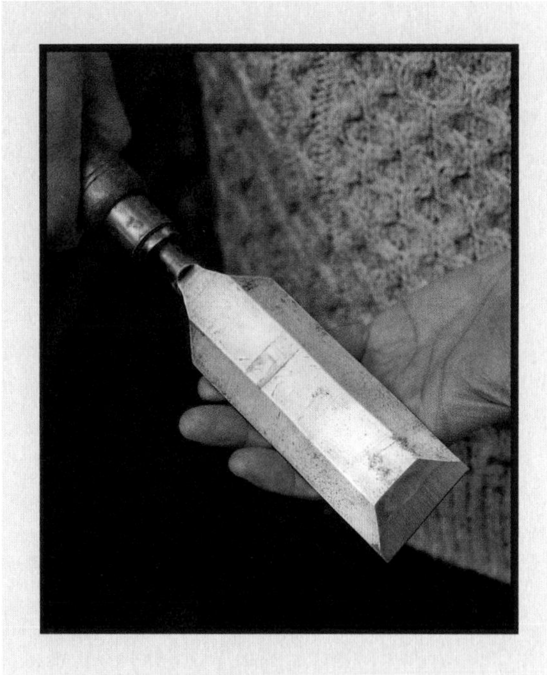

MY FAVOURITE TOOL

'I have a lot of chisels. Some of the cheaper chain-store ones I use for rough work, because if they get damaged it's no great loss to replace them. But my absolute favourite is a chisel I bought second-hand at a market in Bath for about a tenner some years ago. I wouldn't be without it. It's weighty, has a wooden handle and a bevel-edged blade made of Sheffield steel. The blade is so wide you can plane with it, and it stays sharp for a long time, unlike cheaper tools, which tend to blunt quite quickly.

Sharpening blades is a subject in itself. We spent the first two weeks at uni learning how to do it properly. I sharpen my chisels on a whetstone, which is also sometimes called an oilstone, because there is a thin layer of oil on the surface to act as a lubricant. It takes ages, but when you've finished the edge is so sharp you could shave the hairs off your arm.'

COLLABORATION

Suffragette Desk

We met Sophia Harwood Murday from Dumbarton and her aunt, Susie Grandfield, from Hampshire, earlier when they brought in a treasured desk that belonged to their ancestor, Una Duval, a well-known suffragette who, in the early twentieth century, campaigned for the rights of women to vote.

The desk is now under the custodianship of Sophia, who was given it by her aunt in 2020, and who will in turn pass it on to her own eldest niece, Elsa, in years to come.

This beautifully-decorated, Edwardian, bureau-style desk may look delicate, but it tells a story of strength and resilience. However, over the years the desk has become rather battered, and while art conservator Lucia Scalisi is to tackle the paintwork, woodwork specialist Will deals with restoring the wood and attending to the damage.

The first thing he does is give the front of the desk a light clean – he can see where the light bounces off areas of ingrained dirt – then he will seal the area with a coat of shellac polish, ready for Lucia to do her retouching to the decorative paint layer. While he cleans, he assesses the areas of damage. Most of the inside of the desk is intact, the only things missing are beautiful decorative knobs on the inside of the drawers. These need to be replaced.

Structurally the satinwood desk is pretty sound, but the long splits on the left side are a bit of an eyesore, he says, and detract attention away from the lovely decorative painting on the surface. It could also be making the desk a little bit more fragile than it should be.

Will decides that the least invasive course of action is to cut up some strips of sycamore and glue these into the crack

This Suffragette Desk tells the story of resilience in the face of adversity and fighting for the equal right to vote.

to hold the split pieces together. There's so much history to the desk, he adds, he doesn't really want to alter anything. Although the sycamore does not look completely identical to the surrounding area, colour-wise, once it's all glued and dried in place, he is confident he can then take off the excess wood and stain it to match the surrounding area.

One thing that does puzzle Will is that he's heard a scratching sound when the top of the desk is opened up. It sounds like wood rubbing against wood, and he is worried that the paintwork on the front is rubbing against the inside of the top of the desk, which could mean some structural realignment is needed.

He is relieved to find that is not actually the case. He pulls out an insert and realises that a piece

SIMPLE REPAIRS

There are a few simple repairs than you can try yourself:

- To repair a shallow scratch, use an unshelled walnut and rub it over the surface using a circular motion and mild pressure. Then rub the area with your fingers so that the walnut oil penetrates into the grain of the wood. Finally, polish with a soft dry cloth.

- For deeper scratches and cracks you can use wood filler, which is sawdust mixed with a glue such as PVA, which will set quickly. Professional restorers like Will also use shellac sticks – the stick is melted into the scratch and, once dry, any excess is chiselled off.

- The remedy for a white ring or watermark on polished wood is mayonnaise. Apply the mayonnaise to the mark and leave it overnight, then wipe it off the next morning. The vinegar in the mayonnaise will draw out the water. Don't attempt this on bare wood, however, or you'll make a greasy mark.

- Gluing something back together is an altogether riskier business and can make it harder for a restorer to fix an item at a later date. It's better to gather up all the bits and put them somewhere safe before you take the item to be repaired.

of wood at the back hasn't been nailed in properly; whenever the back is slid open it rubs against the inside of the desk and makes an alarming noise. He's confident that this should be quite a straightforward fix. He nails the back panel securely to the back of the insert.

Slightly more complex is casting the three missing drawer knobs or pulls. Will is making replacements himself; one will go on the inside, attached to a small drawer, and the other two on the front of the desk, so Sophia can slide up the lid properly once more.

Rather than turning the replacements on his lathe, Will decides to cast them in resin. So using the existing knob as a template, he mixes up resin and pours it into the mould to make three very convincing replicas. Mixing up pigment – adding a tiny bit of white to create a 'milky, yellowy, creamy colour' – and polish, he then applies thin coats to the surface of the new knobs to colour-match them perfectly into the originals.

Una's delighted descendants are absolutely thrilled and can't believe how beautiful the desk now looks. Susie thinks her grandmother would be 'absolutely over the moon', and can visualise her sitting there with her pen, ready to fight once more for what she believed in.

Will cleans the desk and fixes the long split along the side of the desk, ensuring that the fix blends in seamlessly with the rest of the wood.

Matchstick Clock

A tall order for the *Repair Shop* team

Charlotte Fisher arrives in the Barn with a very special 'grandfather clock'. Crafted from 5,520 matchsticks, the clock, a model of Big Ben (the Elizabeth Tower) in Westminster, London, was built by her own great-grandfather, Edwin Aldous, a model-maker and artist renowned for creating matchstick models of well-known landmarks.

Charlotte, 25, who lives in Bedfordshire, knows all the factual information because Edwin carefully noted it all down on the back of the clock.

Her great-grandfather started making the clock in 1953 to commemorate the late Queen Elizabeth II's Coronation, building the matchsticks around the wood frame and carving them into shape.

He meticulously created the clock tower from memory and also by using photos. On the side of the clock, he carefully replicated Her Majesty's head as it appears on the 1953 shilling.

Standing at five feet tall, the clock took 11 painstaking months to complete. In 1956, the finished model went on display at the International Handicrafts, Homecrafts and Hobbies Exhibition at Olympia, in London. It was seen by Prince Philip, the Duke of Edinburgh and generated lots of media coverage.

Sadly, the clock hasn't worked for at least seven years and its mechanism has been removed. Charlotte planned to have the part repaired, but with the pandemic and lockdown restrictions, she never got around to it. The casing also needs repair. Over the years it has shed many matchsticks, which Charlotte and her nan tried to keep safe.

Three years ago, just before her nan passed away, she told Charlotte that she wanted her to be the clock's caretaker in perpetuity. Charlotte says her

This extraordinary matchstick model was made by Charlotte's grandfather, Edwin Aldous, but needs Will's expertise to restore it to its former glory.

'What puts me in the zone is wood-graining. It gives me real pleasure and I can happily do it for hours. A lot of the pieces I work on have missing patches of veneer, so I'll imitate the colour and grain of the original wood.'

Will Kirk

nan wasn't a sentimental woman, but she loved the clock so much because her father had built it.

She kept it in pride of place next to the television. Winding up the clock was one of her favourite things to do. When Charlotte was a toddler, it seemed ginormous to her, but if she was well-behaved her nan would allow her to wind it up too. It would mean so much to Charlotte to have the clock restored and ticking again, taking her back to those times spent in her nan's house.

While horologist Steve Fletcher looks at the clock's workings, Jay takes the casing to woodworker Will so he can work out a way of repairing the missing matchsticks.

Will sets to work revamping the casing; he's never seen anything as large as this made from matchsticks before. First, he gathers the bits that have fallen off and figures out where they need to go. When he's glued these back on, he gives the entire clock a clean and works out how to replace the missing elements.

Steve admits he's envious of Will doing the match work on the case because he used to make things out of matchsticks when he was a child.

While Will works on the wood, Steve works out what is wrong with the clock mechanism. To do this, the first thing he has to do is to let down the power, gently. There's a huge amount of power in the main spring of a clock when it's wound up, Steve explains. If you undo the plates, and everything goes 'whizz bang', it strips the teeth and causes so much damage it can actually destroy a clock's mechanism. So he dismantles the clock to give it a really good clean and to see what he has to repair.

After checking every single tooth of the workings Steve has found three holes on the back plate that need repair. Because the holes have worn, the pivots, part of the clock's mechanism, no longer fit snugly.

The pivots are wobbling around, he explains, so the wheels start rubbing against each other, causing wear on the teeth of the clock, which creates problems. To resolve this, Steve needs to fit new tight-fitting brass plugs, called bushes. He gently files back the first hole, making it round instead of the oval shape it's become, before selecting the right size of bush to push in, repeating the process with the other two holes.

Will, meanwhile, is in awe of the workmanship on the casing. He painstakingly glues all the loose pieces back into place, then he gives the whole thing a clean before replacing the missing matches. When it's dry, he will colour-match old and new.

Steve adds on the pendulum and it strikes. And ticks. After making a few more adjustments, he pops the top back on.

When Charlotte returns and sees – and hears – the clock, she's delighted. She is instantly transported back to her nan's flat in Dunstable, she says, sitting there with her little bowl of custard and bananas. She tells the team that the clock back in working order is like having a piece of her nan back.

Will carefully colour matches the new matchsticks he adds in to the clock casing.

COLLABORATION

Windrush Clock

Time tells a tale

Dorcas Cain and her brother Stephen travelled to the Barn from Birmingham with a cherished reminder of their parents' determination to make a new life in Britain.

They have brought with them a clock that belonged to their parents, Hermann and Keturah Brown, who came to England from the Caribbean island of Antigua as part of the Windrush generation in 1960. The clock is so special to them because it represents their mother's pride in settling with her

family here. She saw the clock in a shop window and told her husband that it was the one for her.

Dorcas and Stephen would be so pleased to see the clock as it would have looked back then, as Dorcas explains that in those days a lot of Caribbean people simply couldn't get credit or a bank loan to furnish their homes. So her parents worked and saved; this clock was one of the first items they bought to furnish their home.

Hermann and Keturah didn't have a house, just a room. When their family started to come along, they moved into two rooms. After that they got their first council house, where they lived until 1981. Eventually, they were able to buy a house that they owned until they both passed away. And the clock sat on top of the cabinet in pride of place, always in the front room, Dorcas recalls.

It used to chime on the hour, she says. It would chime and the family would know that Dad was coming home, so they would go and wash the dishes. It would chime when it was time for bed, and for church too, Stephen adds.

When they couldn't find a British church that welcomed them, the Browns set up their own, with Hermann becoming church leader and pastor for more than forty years.

But Dorcas thinks it's at least a quarter of a century since the clock has chimed. She and her brother would really like to hear it once more, to remind them of their days growing up in a house full of love.

And the restored clock would honour the Windrush generation, who came here with nothing but their dreams, and stuck it out in the face of so much hostility.

This Windrush Clock serves as a special reminder to Dorcas and Stephen of their parents' determination to start a new life in Britain.

Will strips off the varnish to inspect the staining on the wooden clock case.

It goes much deeper than being simply a clock, Dorcas agrees, it's about pride. Having it repaired would mean so much.

The case of the clock is really quite badly damaged and stained, so Will takes a look while horologist Steve Fletcher inspects the inside mechanism. The problem for the lack of chiming could be something simple, such as a worn bearing or a bent lever, but Steve won't know until he takes it apart.

Will gets to work on the badly-stained case. He uses varnish stripper to take the finish off the surface, explaining that he could sand it, but he's not sure how thin the veneer is. Taking an abrasive approach could go straight through to the wood beneath.

He's not too sure what has dripped onto the surface, but it has definitely eaten through the vanish and buried itself in the wood. Will says there is a slight chance that when he has taken off all the varnish off, he might have to use another method to try to draw out the stain.

Meanwhile, Steve is working on the clock's mechanism. He's found that the pivots of the clock are in really good condition. But when he pops them into the plates he finds that there is some wear in the pivot holes. This is all very precise; he says that one pivot is moving across probably .25 of a millimetre, but in clock-making terms, this is substantial. Steve files the hole to centre it up, joking that he can do this confidently by eye because he's been doing it for 50 years now, so he's getting quite good at it.

Will has now stripped the old varnish off the casing, but there are still some really dark stains on the left. He decides to use wood bleach to tackle them. In theory, the stain should bleach out quite well, he thinks. The outcome depends on the type of wood being bleached and how deep the stain is.

When the bleach has done its job, he has to neutralise the surface with water. The water will also show him how the surface will look like when polished up. It's the moment of truth, but thankfully, he's delighted with the massive transformation the bleach has made.

Steve's also pleased with the progress he is making on the mechanism. As he trips the chime and strike mechanism, he says that even after all the years in which he has been repairing clocks he still enjoys seeing a mechanism chiming and striking like new. All that remains is to pop the hands back on the clock and set up the hammers so the chimes sound nice and sweet.

This very special clock is ready for Dorcas and Stephen to once again hear the sound of their childhood and remember their beloved parents.

Wilma's Table

Rescuing a family favourite

A small, round, wooden occasional table serving as a reminder of a caring and much-missed mum was a permanent fixture in the council house where Wilma Adams grew up with her two sisters, Carol and Barbara, and their beloved mother, Eunice.

With its varnished top and carved or moulded decorative edges, Wilma thinks the table dates from the 1960s. However, it may have been purchased second-hand, which would make it even older.

It belonged to Wilma's mum, Eunice Annetta Perry, and accompanied the family on every house move. Wilma remembers the table always covered with a hand-sewn cloth made by her eldest sister, Carol.

Eunice had limited resources; as a single parent working night shifts, every penny counted. This table was something small but that was saved for hard. Their mum didn't own much furniture, but she really treasured the few pieces she had.

Eunice was born in the Hanover Parish of Jamaica in 1942. In 1961, at the age of eighteen, she came to England with her younger sister, sixteen-year-old Cynthia, to join their father, Percival Perry, who had arrived a few years prior to work on the roads in Bristol. Tragically, just over four months after they arrived, their father felt unwell at work whilst he was using a pneumatic drill. He was taken to hospital and passed away later that day.

Eunice and her sister didn't know anyone apart from him in England, and losing their beloved father was traumatic. Eunice was very resilient, though, and despite this difficult beginning, she cared for her younger sister.

With ambitions to be a nurse, Eunice started her training, but she fell pregnant and didn't complete it. She went on to spend most of her life working nights as an auxiliary nurse at a hospital looking after young people with disabilities.

Wilma says that Eunice was a devoted mother and grandmother who dedicated her life to raising

This table holds memories of a hard-working, caring mum and its restoration means that it can be passed on.

her family. She brought up her daughters as a single mother. She worked very hard, and because she hadn't been able to complete her own training, she invested in her daughters' education. Encouraging them all to become professionals, she was immensely proud to see Wilma, Carol and Barbara all graduate with university degrees. Wilma followed her mother's footsteps and went into nursing.

Eunice was fortunate to live long enough to meet all her grandchildren, taking early retirement to look after her grandson Uriel, Wilma's nephew, who was born with a disability, caring for him 24/7. She also cared for others in her community.

When Eunice passed away suddenly in 2010 following a brain haemorrhage, Wilma was overwhelmed by the messages and stories about her mum that she heard at her funeral.

Wilma inherited the table. Despite the imperfections it has gathered over the years, it has always been on display in her living room at home. Although she is proud of this modest piece of furniture – because it reminds her of where her mum started out and how far the family has come – Wilma would really like everyone who sees it to admire it, because, for her, this would do true justice to the memory of Eunice.

Her sisters will also be so pleased to see the table restored, so that it can then be passed down to any of Wilma's three sons in the future. It would have meant so much to Eunice to know that it is being treasured.

HOW TO CARE FOR WOOD

- Wood needs protecting from excessive heat and light, so it's best to site furniture away from hot radiators and direct sunlight, if possible. Sunlight will fade finishes, so if you place a bowl on a polished side table, for example, and display it in the window, before long you're going to have a dark circle in the middle.

- Always use placemats or coasters to protect wood from spills and rings.

- Polish wooden surfaces with beeswax rather than aerosol polishes.

There's water damage on the top of the table, and some veneer is missing in the centre. The table edge and the shelf underneath are damaged, and there are several chips on the legs.

Woodwork expert Will Kirk's plan is to first clean the table, then glue down any lifting pieces of veneer. He will need to remove the thin varnish layer to get a proper match for the missing veneer.

The challenge is to find a good match for the veneer that's the right grain and colour. Once he has done this, Will patches in the missing veneer on the table top so that the water damage is no longer even visible. His next job is to glue in a new piece of wood where the moulding is missing. He leaves this overnight to dry, then he carefully hand carves the new moulding to match the old.

When this is complete, he polishes the top to make it gleam once more. The legs are given a light sand to remove the scratches, followed by a polish. Will says that these simple steps will all add up to a really nice transformation for Wilma's table.

Woodwork expert Will patches in the missing veneer to cover up the damage, transforming this heirloom for generations to come.

BOOKBINDING

BOOKBINDING
Christopher Shaw

The only day in the year when you won't find bookbinder Chris Shaw engrossed in his work is Boxing Day. That's when he's Morris dancing. The rest of the time he's happily engaged in what he does best, restoring, repairing and binding printed books and other volumes, either at his home studio or in *The Repair Shop* Barn. He'll even sneak in a couple of hours on Christmas Day.

What drew Chris to the craft was his love of books – as tangible objects as much as the texts they contain. In fact, the only thing capable of distracting him from the job at hand is when his attention is snagged by a line of print and he finds himself caught up in reading the pages he is binding or restoring, especially if there are intriguing handwritten annotations in the margins.

Along with a love of books, Chris has a passion for paper, which he shares with his wife Sarah, who also trained as a bookbinder. For years, they've both been avid collectors of vintage, handmade and decorative papers, which are scattered all over the house, tucked away in drawers and cupboards, and even under the bed. Sometimes parting with them is a wrench.

Unlike pottery, which has a history measured in millennia, the craft of bookbinding is a relative newcomer. In its present form, it dates back to the fifteenth century, when the printing press was invented and the technique of rag papermaking arrived from the East. Nevertheless, some of the basic elements and methods involved are much older and have ancient origins. Scrolls and tablets made of wax or clay were early methods of recording information. But book-like objects can be traced back to the 'codices' of ancient Rome and Egypt, where individual leaves of wood, vellum or papyrus were laced or corded together to form 'pages' that could be turned.

The earliest existing example of European bookbinding and the oldest intact European book is the St Cuthbert Gospel, from around 700 CE, which is in the British Library. Its pages are made of folded and sewn parchment and its covers from thin birch boards covered with red goatskin.

Over the centuries, the actual craft of bookbinding hasn't changed very much. That's because the design of a book as a physical object is hard to improve: it's portable, shelve-able, and relatively durable, and it presents a text in an accessible and readable way. Nevertheless, materials have varied. When parchment (or vellum) gave way to lighter-weight paper, heavy wooden boards were no longer needed to keep the pages in place and pasteboard was used to make covers instead. In pre-Victorian times the sections of books were sewn onto strips of hemp. Nowadays the tapes are made of linen.

With mechanisation, commercial publishing has pushed hand-bookbinding to the margins. The fact

that the craft survives today at all is largely down to the demand for handbound copies of postgraduate and doctoral theses, which are the bread and butter of binderies up and down the country, along with presentation editions. Chris, who has been a Fellow of Designer Bookbinders since 2004, has had many prestigious commissions. As well as working on valuable historical books, he's also produced specially bound copies for Booker prize-winning authors and for notable illustrators such as Ronald Searle. His forty years of experience in the craft amounts to a lifelong dialogue with materials, tools and techniques.

The books that come into the Barn present Chris with a wide range of challenges; in some instances he draws on the expertise of Louise Drover, the paper conservator. Rips, tears and loose pages are common. Stapled or spiral-bound notebooks and leaflets can show rust stains where the metal has aged with time, and eraser crumbs rubbed gently along the grain of soiled paper can help to remove the oils that come from handling. Once the item is handed back to its grateful owners, he knows he has done everything he can to protect it, from backing frail pages with tissue, to adding extra linen to the spine to strengthen it. Often, he can't resist providing an extra layer of security in the form of a tailor-made slipcase.

Like all the experts on *The Repair Shop*, Chris's main aim is to extend the longevity of the family treasures that he restores, whether this is an old Bible handed down the generations, a wedding album, or a precious notebook of recipes. By using natural materials and traditional methods he knows that if an item falls apart in years to come someone will be able to repair it again. At the same time, he's not after the kind of perfection that erases all signs of past history and handling: for Chris, every book tells its own story.

Chris sees a wide range of challenges in the books he repairs in the Barn, but is always careful to maintain the history of each book he restores.

Anatomy book

A treasured family heirloom

General Practitioner Dr Ramesh Bhatt was left a beautiful anatomy book by his 'second mum' Margo twenty years ago. He would love to see it restored so that he can enjoy with it his daughter Kavita, who has recently qualified as a GP herself.

Margo (Margaret Epstein) was the mother of Ramesh's schoolfriend Roger. She took Ramesh under her wing when he arrived in North London from Kampala, Uganda, in 1968, at the age of seventeen, to study for O Levels and A Levels.

Ramesh had left his own parents behind – they came to England in 1972 when dictator Idi Amin ordered the expulsion of Ugandan Asians – and joined his brother, Ashok, already a university student in London.

Full of youthful confidence, Ramesh marched into the local grammar school, Christ's College Finchley, and impressed the headteacher so much with his academic curiosity that he was given a place.

Margo invited Ramesh home for dinner, introducing him to a welcoming home filled with books and classical music. Ramesh and Roger became lifelong friends. When Margo died in 2003, she left Ramesh this prized heirloom, believed to be almost 200 years old, in her will.

Growing up in East Africa, and hearing stories about famous doctors from the neighbouring Congo, Ramesh knew from an early age that his future was in medicine, and he 'fell in love' with the idea of the NHS. He qualified in general practice at UCL (University College London) medical school.

Ramesh shares that ironically, he 'never was very good at anatomy', his forte being general practice. This volume reminds him of what he doesn't know, and he still contemplates the illustrations for inspiration. For modern clinicians like Kavita, the book shines a light on how medicine was taught and perceived in the past. When she compares the volume to her own student anatomy books, she can't believe the level of detail.

It holds huge emotional significance too. Kavita remembers Margo, and she now lives opposite Roger and his wife Helen, who are her godparents. Kavita can't imagine the book not being there. It's part of the family and something she is very proud of, both for its link to Margo and Roger, and its significance in the history of medicine.

Sadly, Ramesh and Kavita are now very wary about opening the book because it has become so delicate and they worry it will only become more damaged over time. Out of respect for Margo, Ramesh would like to the book in better condition to pass to Kavita.

The leather-cornered board covers have come off, the spine is in pieces, colours are faded, and some leaves are loose. Father and daughter are very keen for the book to look as close to the original as possible, asking expert bookbinder Chris Shaw to restore the colours of the spine and cover, which have become faded.

Minor repairs with heat-sensitive Japanese tissue and careful ironing to bond old and new fixes the loose leaves. Chris decides that the book definitely requires a new leather spine and the leather corners restoring. He creates a new spine and adds the publisher's details by heating up the alphabet tools, pressing them into the leather, then laying gold leaf into the indented letters.

He also completes a full re-bind, using a centuries-old technique of threading hemp laces through the covers, hammering down the hemp to ensure the binding of this treasured textbook is as secure as possible from now on.

This Anatomy Book holds special memories of a 'second mum' for General Practitioner Dr Ramesh Bhatt.

British Passport

Telling a story of British-Jamaican identity

A prized memento belonging to Edgar Alfanso Whyte, known to all as Alfanso, born in Port Antonio, Jamaica, in 1926, is brought to the Barn by his daughter, Beverley Dixon, from London. It's a British passport, issued to her dad in 1948, when Jamaica was still a British colony. Like many young West Indian men of the time, Alfanso answered the call from the 'Mother Country' to help rebuild Britain, travelling on the SS *Eros*, a journey that took three weeks in 1950 as part of the Windrush generation. As a young man, Alfanso settled in south London and followed his trade, tailoring.

The cherished memories that his official identity document hold are very special to both Beverley and Alfanso, who is ninety-seven and often finds it hard to recall his past. Beverley feels that if his passport could be restored to it's original pristine condition, it might help him remember this important time in personal and social history. Being able to hold his repaired passport, its gold crest glimmering once again, she hopes hopefully would prompt discussion around Alfanso's journey.

Beverley says that without the passport her dad wouldn't have been able to build a life in Britain, and she wants to celebrate the fact that he and her mum created a strong foundation here for their family.

Alfanso lost his mum and sister around the age of twelve, so he forged his own way with the support of his dad, who owned a bakery business, and extended family.

Alfanso met Beverley's mum, Rubena, in Jamaica; she was training to be a seamstress specialising in embroidery. After he arrived in Britain, he sent for her to join him – she arrived on the SS *Colombie* in 1951, and they married at St John's church in Brixton, south London, in 1952.

A British passport was a prized possession in the Jamaican community, Beverley explains. And now its story is part of the Caribbean's oral history, as it represents Alfanso's journey to Britain as part of the historically significant Windrush generation. She

This official document holds cherished memories of building a new life in Britain and Beverley hopes to spark conversations with her father about this important journey.

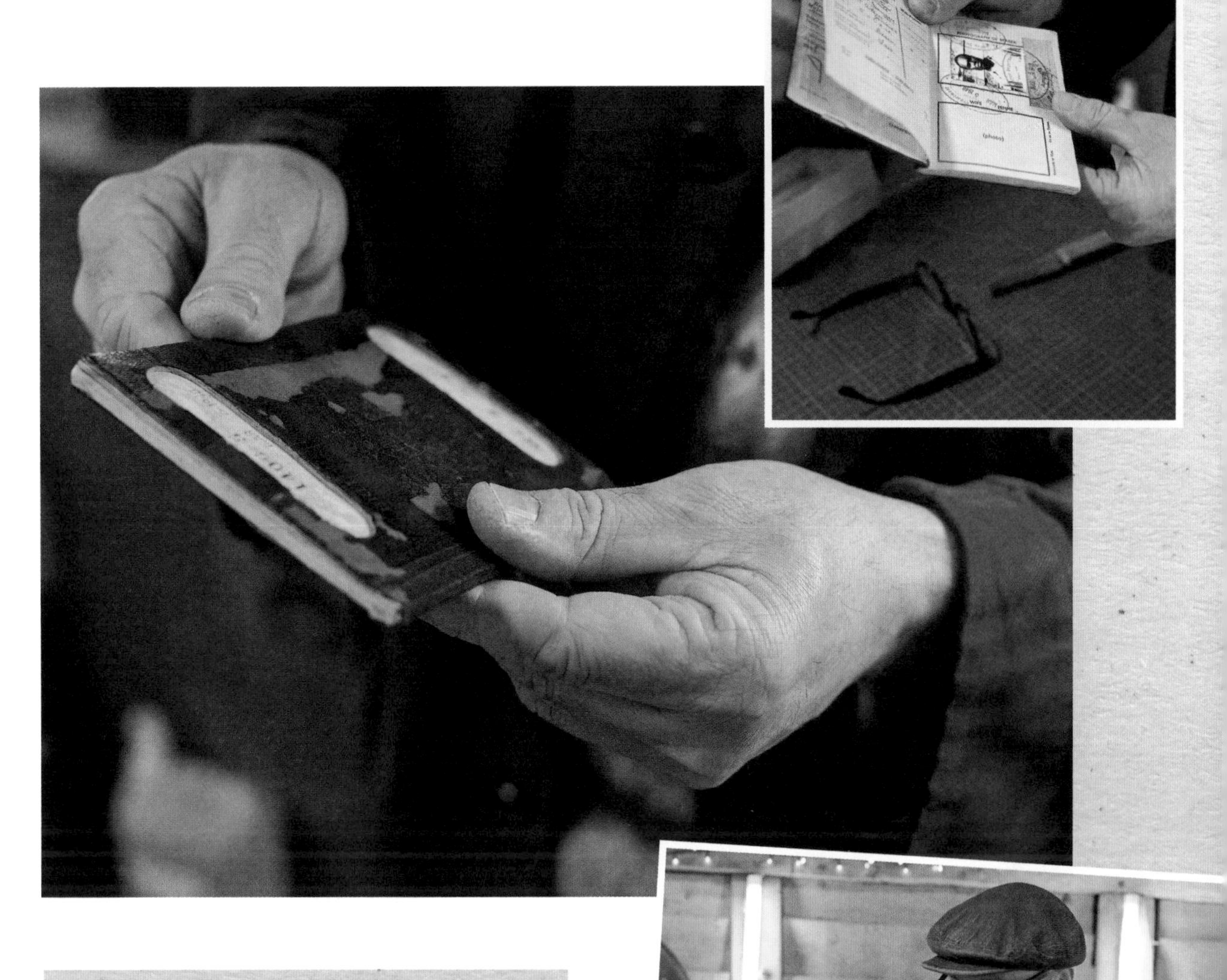

'Mindfulness is involved in every single stage of bookbinding. For me the sweet spot is sewing. I can work at a relaxed pace while listening to Radio Four.'

Chris Shaw

feels very strongly that such stories and memories should be recorded before this older generation pass on. For Beverley, this restoration is a way of giving back to her father's generation, who sacrificed everything to give their families a better life.

It would also be a significant gesture for the next generations. Alfanso has four children, eight grandchildren and, four great-grandchildren, with one more on the way. Beverley shares that the younger family members are fascinated by the passport, and like to hold it, imagining themselves as young children in Jamaica, or spending three weeks on a ship.

There is also a serious message. Many questions are asked about whether people from the Caribbean are British, Beverley says. For her, it would be great to have her father's passport restored as that documentary proof of him being a British subject and Jamaican, which was of great importance at the time. The Caribbean country became independent with full dominion status within the Commonwealth in 1962, under a constitution that retained the British monarch as head of state.

The passport has been well looked after for nearly eighty years. However, it is now showing signs of its age.

Bookbinder Chris Shaw is ready to get to work. He's never restored a passport before, so is fascinated to learn how it is constructed. It's coming apart because it's suffered from damp, he says, which has

affected the glue and has caused creasing damage to the pages. To repair the passport, he needs to separate the cover from the pages and reinforce it with new board. He also selects some new leather-effect cloth as a covering.

He presses all the layers with blotting paper to start with, pops it into the nipping press and leaves it as long as possible so that they all stick together firmly, without creases.

There is a black residue obscuring the gold crest on the front of the passport. Using water and a scalpel, Chris gently lifts it all away.

He notices that the sewing holding together the pages has become exposed, causing them to come loose. So he strengthens the stitches with tiny little pieces of Japanese tissue along its edge to help support the paper; this will prevent the pages falling out in future. Then he re-inks Alfanso's name and passport number, so it looks as strong and clear as it did the day he left Jamaica.

As a final touch, Chris makes a special protective box folder for Beverley to keep her father's passport in from now on, complete with a slot for her mother's passport too, which is embossed with the names of the ships they travelled on to Britain to make their new life, SS *Eros* and SS *Colombie.*

Restoring this British Passport is a first for bookbinder Chris, who has never restored this type of document before.

MY FAVOURITE TOOL

'My grandfather was a leatherworker and I now own a number of tools that used to belong to him. One of these is a bone folder, which must be eighty years old. A bone folder isn't just useful for folding. You can apply pressure with it and use the tip to close up sewing holes to stop the glue seeping down on the pages when you attach the sewn sections to the spine. I use a bone folder every single day and there are three that I am really fond of. They sit so well in the hand and are very personal to me – the smoothness of the shape and fit is beyond physical, it's almost an instinctive connection, like Harry Potter's magic wand.

When one my previous bone folders cracked after 30 years' use, it took me fifteen years to get truly comfortable with its replacement!'

Michael Rosen's Patient's Diary

Poignant Covid-19 memories from a treasured poet and author

A very special visitor to the Barn arrives with a big challenge for bookbinder Chris Shaw. It's much-loved children's author, poet and former Children's Laureate Michael Rosen, and his daughter Elsie. They have brought with them two very special projects that they need help with; a daily diary kept by nurses and carers for Michael whilst he spent forty days in an induced coma in an ICU (intensive care unit) at the Whittington Hospital, north London, at the height of the pandemic, and a bundle of letters, postcards and drawings sent to Michael by children across the UK, wishing him well and hoping for his recovery.

The 2020 diary is in the form of a modern spiral notebook with perforated pages, filled with daily notes and memos written by nurses and carers. At the end of their shifts, they would write a summary of the day and what care Michael was given. This was particularly pertinent as strict Covid-19 restrictions meant that no family were able to visit, so staff members were the only people Michael had contact with. It touches Michael to know from the diary that even though they must have been so concerned about what Covid might bring, the nurses were looking after him, noting when they held his hand, or sang to him.

This 'Patient's Diary' was given to Michael when he was discharged from the hospital's critical care unit, after spending a total of forty-eight days battling the virus.

Michael Rosen is full of admiration for how the NHS dealt with the Covid-19 pandemic.

His record of the time he spent in ICU is important because patients who are as seriously ill as he was often have little or no memory of their stay, as they can be so affected by their illness and the sedative drugs they are administered. To help people understand what they have been through, many NHS hospitals have introduced Patient Diaries to detail daily updates about procedures carried out and patients' progress.

Michael, who has delighted generations of children with books such as *We're Going on a Bear Hunt*, started to feel ill with what felt like flu at the start of the pandemic, in March 2020. He was sick for around two weeks, but then he started to feel worse. His wife, Emma, noticed that he couldn't catch his breath. Elsie says that seeing her dad like this was terrifying; she and her mum feared for the worst. They had no idea what to expect. Michael

was taken straight to ICU. The last conversation he remembers was a nurse or doctor saying, 'will you sign this piece of paper to allow us to put you to sleep?' When he asked if he would wake up, they said he had a '50/50 chance', Michael recalls. When he asked what his chances of survival might be if he didn't sign, they said 'zero' – so he signed. An induced coma puts the brain into hibernation, allowing the body time to repair. When Michael was taken off the induced coma medication, he was finding it difficult to come round, so his consultant, Professor Hugh Montgomery, asked Emma to help. She played recordings of Michael's children in his ear. This tactic was declared a 'gamechanger' by Hugh. Michael woke up.

Michael couldn't bear to read the diary for a year after he left hospital, finding the prospect too terrifying and traumatic. Every time he tried opening and reading it, he was overwhelmed by the love and compassion shown by the people who looked after him. He is full of admiration for how the NHS responded to Covid-19, the biggest challenge in its seventy-five-year history, and the care shown by everyone who stepped up and effectively saved his life. However, he is worried that the notebook itself will not stand the test of time. Michael is now looking to the future, but his diary and the letters are part of a hugely significant part of his past – for him and his family. And they are also a legacy of a time in history when the whole world stopped, except for those on the NHS frontline.

Chris uses decorative papers from his own collection to craft a rainbow for the cover of the scrapbook album.

Michael would dearly love these precious items to become something tangible for future generations to understand. One entry page of the diary has already fallen out. Several other pages are becoming loose. The children's letters and drawings are simply in a bundle and at risk of being damaged.

However, Michael is keen to stress that he doesn't want the diary to change in any way; he would like the page that has come out to be placed back and for the remaining pages can be strengthened to prevent them from falling out in the future.

Chris thinks that this is such a heartwarming project and is keen to see what he can do to help. He says it is good that the diary is in a ring binder as it will have some 'give' to allow for any additional thickness from the heat-sensitive repair tissue he is planning to add.

The problem is obvious; the diary's pages are perforated, so they are easily tearing away from the metal spiral. So Chris add strips of strong heat-sensitive Japanese tissue paper which seals and strengthens the notepaper so Michael will be able to turn each one without fear. Chris cuts the Japanese tissue carefully to size. Taking a heated spatula, which activates the glue, he rubs along each page to join the two elements together. It's a very delicate process; he says he almost feels like a surgeon with a very precise tool.

For the scrapbook album to hold the children's letters and cards, Chris decides to create it in the form of a rainbow, to echo the symbol of hope and public support for the NHS from the pandemic. The pages are double-sided and hinged, which will make the album very useable.

Chris selects a series of decorative papers from his own collection, which he sews together, enjoying the calm and mindful activity. He has selected papers in colours as close to a rainbow as he could get. He admits it was difficult to get the exact match of violet and indigo; the violet one is a paper made of printers' waste, dating from the 1870s. The green paper Chris made himself at college, more than 40 years ago and has never used it. Now he has found the perfect project, he says. He hopes that when Michael takes this very special scrapbook home, he will have as much enjoyment studying it as he has had making it.

When Michael returns, he is absolutely delighted with both the scrapbook and the diary. Amazed to see the pages of the diary now invisibly attached, he says he is overwhelmed to see something so beautifully made, 'complete and tidy and there for me whenever I want to look at, which will be quite often, believe you me.'

Michael is overjoyed to see his diary and scrapbook carefully preserved for him to remember the wonderful NHS staff and people who helped his recovery.

BINDING

Bookbinding encompasses a range of skills, from measuring and cutting, to folding, sewing and gluing. Each demands precision and a steady hand. If you make a mistake at any stage, it can't be covered up and will adversely affect the end result.

The first step is folding, which is where the bone folder comes in. A rounded flat shape made of bone, it's one of the bookbinder's most essential tools. To create sections of the book, or 'signatures', pages are folded in half three times, making eight leaves or pages. Paper must be folded so that the direction of its grain aligns with the spine. To prevent bunching at the corners on the third and final fold, bookbinders will generally slit along the top of the previous two folds almost to the end so that the section lies flat. Alternatively, individual sheets can be folded in half and interleaved.

When all the sections are complete, an even number of holes are pierced down the centre of each one using a pointed tool such as an awl. Then the sections are sewn together using a needle and linen thread. Some bookbinders wax the length of the thread with beeswax so that it slips through the holes more easily.

There are different ways of creating sewn bindings. Stab binding, for example, is a simple Japanese technique that does not require the papers to be folded but secures them with over-stitching. Chris takes the more traditional route and sews folded signatures onto linen tape so that they can be bound into hard covers.

To secure the text block into its covers, Chris uses a specially formulated wheat starch paste as adhesive. Preferred over animal glue by conservators of paper artefacts, wheat starch paste forms a strong, reliable bond and is similar to Japanese nori adhesive, which is made of rice starch.

Similarly, the paper Chris uses in practically every book repair is Japanese, which is available in a huge range of weights and textures. The thinner tissues are ideal for backing old paper to stabilise it. Japanese tissues, with their long fibres, are very strong. Applying a little water allows them to be torn into shape. You can either use a fine-pointed brush to draw the required shape before tearing, or brush a little water down the edge of a ruler for a straight edge. Either way, the torn edge will be feathered, which means it blends it more easily with the original paper.

Book covers have an obvious function: to protect the pages from discolouration and tearing. But they are also the vehicle for further embellishment. For devotees of decorative paper like Chris, this is where covering boards with a beautiful marbled or handmade paper can really lift a book out of the ordinary. Alternative traditional materials are fabric, such as linen, and, most luxurious of all, leather.

For making leather bindings, Chris sources high-quality naturally tanned goatskin. The wonderful grain of the leather is perfect for the ultimate decoration: gold tooling.

GOLD TOOLING

Gold tooling is one of the most meticulous of all the activities associated with bookbinding. The layout has to be absolutely accurate and precisely aligned: you don't get a second chance. While Chris finds the technique absorbing, he admits it can be extremely demanding and time-consuming.

First, an indentation is made in the leather with a heated metal-tipped tool. Then adhesive – water-based 'glaire' made of egg albumen – is applied to the blind impression and left to dry, or cure. Gold leaf is subsequently laid on top and held in place before the adhesive is reactivated using the same heated tool to glue the gold in place permanently.

The tools themselves have a range of profiles, from alphabets in various different typefaces, to lines, ornamental flourishes and curlicues. Once, when a bindery in Northampton was closing down for business, Chris bought up a collection of their tools. Years later, when a Victorian Bible came into his workshop for repair, held together with gaffer tape and with part of its spine missing, he realised that he had just the right tools to replicate the missing lettering – not surprisingly, since he discovered that the Bible had originally been bound at the very same Northampton bindery where he had acquired the tools.

METALWORK

METALWORK

Dominic Chinea

With a background in classic car restoration, along with set building and design, Dom Chinea, *The Repair Shop*'s resident metalworker, can turn his hand to a wide range of challenges. The breadth of his skills and his determination to keep learning new ones means that he often works in collaboration with other experts in the Barn.

Today, metalwork is not as popular or as familiar a craft as it was in the days when it was widely taught in most secondary schools. In fact, some skills, such as bell founding and coppersmithing, are in danger of dying out completely. Recently added to the Red List of Endangered Crafts is wheeling, something that is close to Dom's heart.

Wheeling is a way of making compound curves, such as those used in a car's wheel arches, for example, by rolling a flat sheet of metal through a special tool to manipulate, stretch or shape it. The tool itself is a large cast-iron frame, like a closed 'C' or giant vice, with a rolling wheel and an anvil wheel at its centre, through which the metal is fed back and forth. It takes a lot of experience to perfect the technique – the slightest movement will affect the result. Wheeling machines are sometimes known as 'Spitfire' wheels, because they were used to make the fuselages and wings of those fighter aircraft during the Second World War. As such, they are a key element of Britain's industrial heritage.

To help keep these skills alive and bring them to the attention of the next generation of school-leavers, Dom spreads the word through the activities of bodies such as the Heritage Crafts Association, who run open days, and the Heritage Skills Academy. The Heritage Skills Academy, based in Bicester and Brooklands, recruits and trains apprentices for careers in automotive engineering, particularly classic car restoration. Apprentices literally learn on the job, taking up paid placements with leading manufacturers such as Aston Martin and Bugatti. Such schemes help to solve a central dilemma of those keen to take up a craft, which is how to earn a living from your passion.

Dom, who classifies himself as a maker and restorer rather than someone who practises a craft, tends to get objects to fix and repair that once had a particular function, rather than those that are simply decorative. This adds a layer of complexity to his work, particularly when the object has been homemade from salvaged parts. Those repairs may be the most challenging, but they are also the most rewarding. The key, he says, is to 'work smarter, not harder'. It's not about brute strength.

The first stage in any project is to puzzle out how to proceed, and Dom is careful not to rush before he's got a clear plan of action mapped out in his head. Part of this process is establishing exactly what metal he is dealing with: different types have

different melting points, which will determine the best welding or jointing method.

The next stage is generally disassembly, which presents the first struggle, since fixing components such as bolts, nuts and screws are often rusty and stuck in place. You need to be systematic, numbering and labelling pieces, if necessary, and noting the order in which they were put together. At this point, it usually becomes clear if someone has attempted a previous repair, if components are missing or in the wrong position. Occasionally, an missing element may have to be fabricated from scratch.

There are various methods of removing rust, grease and accumulated dirt. Small rusty items can be treated with rust-removing solutions or by electrolysis; larger ones are often sent off for sand-blasting. Otherwise, it's a question of using either wire wool or wet and dry sandpaper. Both these are available in different grades of abrasion.

Working with metal entails learning a broad range of skills, from cutting with a fine-toothed metal saw or angle grinder, to hammering, panel beating, drilling and welding. Protective goggles and a mask are essential when carrying out any task that might cause sparks to fly or give rise to toxic fumes.

For welding steel, Dom uses a MIG welder, while for welding aluminium, brass or copper he uses a TIG welder. Brazing, which is a form of brass welding, Dom uses less rarely. Since he's been working at *The Repair Shop*, he has added soldering to his repertoire, picking up the skill from silversmith Brendon West and instrument repairer Pete Wood.

Like many makers and restorers, Dom has a huge collection of components – from nuts and bolts to metal parts salvaged from other objects. If he's working on a repair where an element is missing and he can't find a suitable substitute, he'll make it himself.

This can-do approach is one of the reasons why Dom is so eager to keep learning, whether he picks up new skills from his colleagues on *The Repair Shop* or by enrolling on courses. Lockdown fostered his interest in gardening and beekeeping and more recently he's learned how to gild and reverse-gild. His advice is to always give things a go and see where it takes you.

Working with metal uses a broad range of skills and tools.

TYPES OF METAL

The most common types of metal, apart from gold and silver, which are in a league apart, are copper, various kinds of steel, iron, copper, chromium, aluminium, tin, brass and bronze. Most can be bent, cut, heated and shaped, and many are shiny and reflective.

- **Tin** is a silvery metal with a relatively low melting point. Today, it's most commonly used as a plating for steel to prevent corrosion, or to make alloys such as pewter and bronze. Cornwall was once a leading centre of tin-mining.

- **Copper**, like tin, is a metal that has been worked for millennia, used to fashion a wide range of artefacts. Naturally a warm reddish colour, it acquires a green patina with age. Softish, and a good conductor of heat, today it is most frequently used in wiring, roofing, currency, plumbing and to make pots and pans.

- **Bronze**, a metal formed by smelting tin and copper, has given its name to a whole period of prehistory, which immediately preceded the Iron Age. It was probably discovered by accident. More durable and stronger than copper, it's been used to make a wide range of products and artefacts, from tools to bearings and springs. Because it's resistant to salt water corrosion, it's often used to make boat propellors. As the name suggests, bell metal, a bronze alloy made with a high percentage of tin, is used to make bells, while since the earliest times statuary and other decorative artefacts have often been made of cast bronze.

- **Iron** has a high melting point and is very soft in its purest form. Treated in a blast furnace it becomes hard and functional. Molten iron poured into a mould and allowed to cool creates cast iron, which has a higher carbon content than wrought iron. Wrought iron, the product of molten iron and slag (a waste by-product of its production), is tough and malleable by hammering, rolling and forging. Once used to make a huge range of items, from swords and cutlery to nails, bolts and tools, the term 'wrought iron' today often refers to mass-produced items made of mild steel, not forged by hand. Iron is prone to rust.

- **Steel** comprises a broad family of materials, created from pig (or raw) iron in combination with other metals to engender different strengths and performances. The most common are stainless steel, tool steel, carbon steel and mild steel. Stainless steel, which resists corrosion, is a blend of pig iron, chromium and nickel. Tool steel, hardened by tempering, is pig iron blended with nickel and tungsten. Carbon steel is prone to rust and difficult to work, while mild steel has a relatively low carbon content. Chromium plating applied to steel is a familiar element of car parts and trim. With a host of applications, steel is the workhorse of many manufacturing industries, from construction to the automotive industry.

- **Brass** is an alloy of copper and zinc. Strong and highly versatile, it has a wealth of applications from door knobs, screws, fittings and fixtures, to tools and wind instruments. With time, its bright shiny finish tarnishes, but can be easily restored using brass polish.

- **Aluminium** is a relative newcomer and was only discovered in the nineteenth century. Silvery-white, three times lighter than iron, resistant to corrosion, extremely pliable and easy to recycle without any loss of quality, it has a broad range of applications in construction, aviation and the automotive industry.

B.G.B
MOTOR

HOT
TOWELS

Hot Towel Machine

Vintage barbershop favourite

An elaborate contraption holding cherished memories of a much-missed father is the most treasured item Panayides Adamou, known as Panos, has in his barbershop on the Isle of Wight. It's a 1950s-built chrome-plated hot towel machine, once a fixture in barbershops everywhere.

However, the machine no longer works and has become tarnished over time, falling into disrepair when customers stopped requesting wet shaves. It's always a talking point, but it hasn't operated for almost three decades. Panos would dearly love it to work once more, and for the outside to be rebuffed to its original shiny finish. Then he will be able to offer wet shaves again, as a luxury old-school service for his customers.

Hot towel machines are a rarity nowadays and this treasured item offers Panos the opportunity to offer an old-school wet shave to his customers.

Hot towels are essential for cleaning the skin's pores during a beard trim or a face shave. These days, Panos explains, hot towel machines are a rarity; microwaveable pre-packed towels are used instead.

Dom explains that a hot towel machine works like a kettle; it is plugged in, filled up with water, brought to a boil and then it steams the hot towels inside an oven-like compartment, making them soft and luxurious.

During this traditional and elaborate way of doing things, the barber will always put on a grand show as they unfurl the towels and get to work.

There's a fascinating family story behind this hot towel machine too. It originally belonged to Panos's dad, Peter. He was working at a barbershop in Covent Garden, central London, when, in 1975, the owner decided to retire and gifted to him. It's been in the Adamou family ever since.

And now Panos's son, Alexander Panayoitis Adamou, has followed his dad into barbering – that's three generations in the same business. Family heritage is very important to Panos. His father was born in 1937 in Limnia, a small village in Cyprus. Panos says it was a hard upbringing; his dad's family had little money and there was limited schooling in the area, so young people had to go straight into jobs.

Peter and his younger brother Vassillis Adamou, known as Vasos, emigrated to the UK around 1960 when Vasos was just twenty years old. With few employment opportunities in Cyprus, they wanted to come to Britain to work hard and make money.

Settling in Archway, north London, within a growing Cypriot community, Peter ended up as an apprentice for an uncle, who was already a barber. There were no training courses, you had to just

Metalworker Dom descales the hot towel machine and fixes the holes, so that this treasured object can work once again.

look and learn, Panos says. His dad was thrown in at the deep end, giving haircuts and shaving under instruction until he gained experience in various barbershops around the capital.

During this time, Peter met his wife, Pinelopie Kyriacou Pantelli, and they had three sons, with Panos being the middle child, born in 1965. The same year, Peter opened his own barbershop on Oxford Street, central London.

In 1973, Peter, his wife and children, including an eight-year-old Panos, moved back to Cyprus. But only a few months later Turkey invaded, and the family had to be evacuated back to the UK. Peter had no choice but to find work in a factory, until he heard about an 'old boy', in Panos's words, who needed a hand in his barbershop. Peter started working for him in Covent Garden, inheriting the hot towel machine and taking over the business when his boss retired in 1975.

Peter worked here until he retired himself. It's also where Panos and his brothers learned their trade; two of them still run this barbershop today.

Panos says barbering was a popular career in the Cypriot community because reading and writing English wasn't a huge requirement, just skill with the scissors. Just as today, he explains, Cypriot immigrants influenced British culture in so many ways. Barbering as a skill rose in the community, and became a symbol of trust, a helping hand for fellow Cypriots to progress.

The hot towel machine had always been in the family, but Panos only found out about it when he was twenty-one and started his hairdressing apprenticeship. Trying to master the towel machine under his dad's watchful eye wasn't without mishap. One day, after finishing a beard trim, he pulled out a towel and forgot to shake it to cool off the steam. His father was aghast, telling him off in Greek. But despite the searing heat, the client was delighted and said it was the best hot towel he had ever had.

MY FAVOURITE TOOL

'My favourite tool is not a tool as such, it's an old engineer's box that I picked up at Beaulieu, and you can usually see it in the background when I'm being filmed. It's made of wood and has lots of narrow sliding drawers where I keep my signwriting brushes, my pigments and other bits and bobs. I'm not the most organised person generally but I like to keep my tools and components in order – you don't want to just sling them in a bag where they will all get jumbled up together. Everything has a home.

Otherwise, my favourite tool is whichever one I need to work on the job in hand. In general, I'm passionate about educating people to invest in old tools, not to automatically buy new ones. Old tools are often higher in quality and were made to last.'

Sadly, Peter passed away in 2016, a slow decline due to frontal lobe dementia that immobilised him for a lengthy time. His classic hot towel machine brings Panos fond memories of working together in the family business.

To have it up and running again would be a dream come true. So talented metalwork expert Dom Chinea takes a look.

Dom finds that the biggest problem is limescale, caused by water and steam over the years. It's so thick it's like chunky, chalky white cement, completely clogging up the workings of the hot towel machine.

The first step is to soak the mechanism and tap in descaling solution for hours. Dom would really like to keep the original tap, but of course it has to work. He spends a lot of time examining the tap and is relieved to find that it can be salvaged.

The insides of the machine have also become riddled with holes, so Dom needs to solder these too. And the aluminium shelf holding the towels has fallen apart, so he makes a new one.

Finally, Dom researches what the original 'hot towels' sign would have looked like in the 1950s. He settles on plain cream glass with smart black lettering. When he's put everything back together again, the hot towel machine looks as good as new.

BLACKSMITHING

Blacksmithing is a type of metalwork that demands a unique skill set and a fair degree of physical strength. It dates back over 4,000 years. One of the heritage crafts taught at the Prince's Foundation craft programme at Dumfries House, the trade is not at present endangered, although current demand for blacksmithing skills is far greater than the number of practitioners available to fulfil it. Jeremy, a graduate of the course and occasional tutor, came down to *The Repair Shop* to help Dom restore a set of fire implements, a job that involved forging a new poker and fire tongs. In his words, the craft can be summarised as: 'get it hot, then hit it'. Working by a roaring fire is not everyone's idea of the perfect working conditions, but there is nowhere else he would rather be.

In blacksmithing, metal, usually wrought iron, is heated to a high temperature in a forge, then shaped by striking it with a hammer on an anvil. To create an opening in the metal – to make a handle, for example – the red-hot metal is hit with a punch rather than drilled. This goes with the grain of the metal so that it flows round the opening to make a continuous path of strength.

WD-40
wd40.co.uk/lifehacks

> 'I classify myself as a maker and restorer, rather than someone who practises a craft. I tend to get objects to fix and repair that once had a particular function, rather than those that are simply decorative.'
>
> **Dom Chinea**

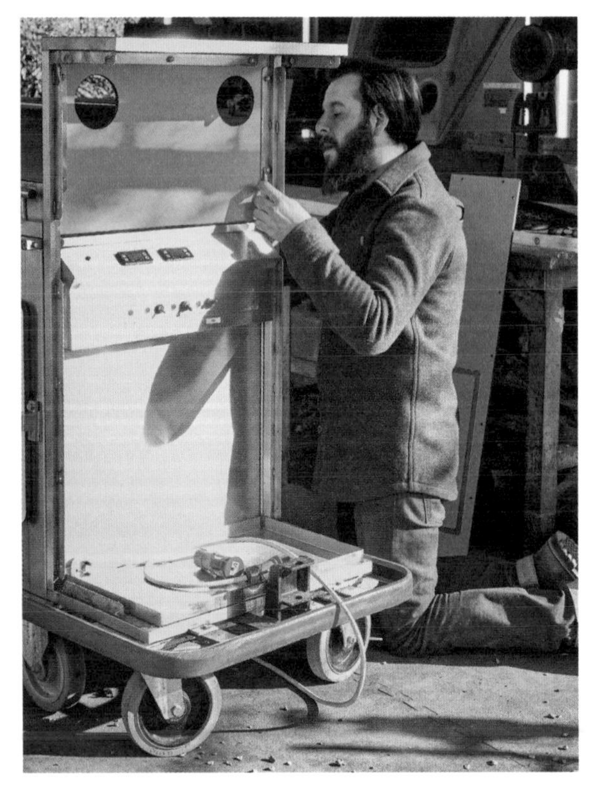

Hospital Trolley

A much-loved hospital fixture

Children's ward senior sister Katie Wilson and her colleague Catherine Read turn up at the Barn with their beloved but battered catering trolley, in constant use every day for breakfast, lunch and dinner at Frimley Park Hospital in Camberley, Surrey.

Nicknamed 'Thomas' after the famous storybook tank engine, the trolley is at least three decades old and showing his age. Catherine, a nurse, has been working on this ward for thirty-one years. Thomas the trolley has been with her all that time, probably arriving before she did.

He's like a beloved member of staff, his cheerful painted face helping Catherine and Katie make their patients – who often suffer from chronic conditions – feel better.

Catherine is coming up to retirement and she would love to have Thomas restored so he can carry on the good work. He's been so busy over the years, he has various things wrong with him. He's scuffed and battered, and his wheels are misaligned, making him difficult to push. Handles have gone missing and been taped over. The blue paint for the hot and cold sections has worn off and the 'number one' in the middle is no longer there. Also, one of the serving trays is missing.

Dom is ready to set to work. It's a real honour to be mending Thomas, he says, and he feels really, really proud to be able to give something back to the NHS. When he was a newborn baby, Dom spent some weeks in intensive care suffering from a stomach problem, meaning he couldn't keep food down. It was a terrifying time for his mum. Without the NHS and children's wards like this one, he wouldn't be here now, he admits.

There are two sides to the repair, Dom explains. The first is 'the practical side' – the trolley does need to work technically. Luckily, to his relief, he finds the electronics are fine. However, the door handles are missing, plus there are lots of bits he needs to fix to get the trolley working mechanically as it should, including the missing shelf and oiling and servicing the wheels.

The other side of this repair is the cosmetics. As the arrival of Thomas is the highlight of the children's day, he needs to look fun and cheery, but he's certainly not looking his best.

Thomas the Hospital Trolley helps bring joy to the children on the ward at Frimley Park Hospital in Camberley, Surrey.

> **'Metalwork involves various skills and you have to keep your wits about you. But what puts me in the zone is only tangentially related to working with metal: it's painting, specifically sign-painting. I can sit and paint and the hours just go by. It's a lovely feeling when it's going well.'**
>
> **Dom Chinea**

Dom re-paints Thomas to bring him back to life, which is one of Dom's favourite parts of restoring objects.

Starting with Thomas's face – so he doesn't have to look him in the eye when he's taking him apart – Dom dismantles the trolley.

His first priority is to fix the handles. First, he welds up all the stress cracks and fractures to strengthen the metal, before adding four new handles.

Dom also crafts a new serving tray. There were originally two trays. It would make the nurses' lives so much easier to have them both.

Using the existing tray as reference, he takes a sheet of stainless steel that's the same thickness and sets about 'a bit of metal origami', using his folding machine. This folds each section individually and then Dom pulls up a lever at the bottom that bends up the metal and pulls it round at the corner. Once all the corners are done, he welds them into place.

After re-painting Thomas's face with a bright cheerful smile, Dom admits that getting this very distinctive trolley back together has not been straightforward. He's relieved to be now bolting on the last few pieces at the end of a very long road – or track.

But Thomas's own journey is far from over. He is personally delivered back to the ward by Dom and Jay, ready to serve up pizza, chips and peas to the excited children.

CROSS D
AUTO
SUPPLIES
OPEN

1960s Scooter

Treasured toy rides again

This beloved pedal scooter was bought by ninety-two-year-old Harry for Belinda, his daughter, who lives in Kent, when she was five years old, in the early 1960s.

Worn out from being played with for six decades, the scooter is definitely in need of repair. Harry says he 'hasn't the foggiest' where he bought it from, but it's been handed down over multiple generations now and he would love it restored so that he can pass it on to his great-grandchildren himself.

Belinda was born in Greenwich, where her dad Harry still lives. She remembers zooming down from the top of General Wolfe Hill in Greenwich Park on the scooter. It was always her and her dad's 'thing' to take it around Greenwich, past the Royal Observatory and the Cutty Sark. When Belinda's boys were young, they would go down the same big hill and see who could get the furthest without peddling.

The next generation to enjoy the scooter was Belinda's two sons, now thirty-eight and thirty-four, and her brother Chris's three children. And now there's another generation waiting to take it for a spin – even though Harry's great-grandchildren are too young to use the scooter right now, he wants it to be ready for them to enjoy when they are big enough. Most scooters now are made of plastic, he says, joking that this family treasure is a real relic from the age of the dinosaurs.

Sadly, by the time the scooter reached his grandson, Harry, this relic was already in a sorry state. It's been so well loved by all the family it's literally falling apart.

The metal is riddled with rust, scratches and dents and one of the metal brackets above the back wheel mudguard has disintegrated. There's rust on

This toy scooter holds special memories of family days out in the park, and Harry hopes to pass it on to his great-grandchildren.

the pedal, and the stand has seized up and no longer goes down.

Once the scooter is fixed, Harry says he'll definitely give it a go again, with both his great-grandchildren in his arms as well. The nonagenarian still enjoys life to the full, living in Greenwich close to his roots, travelling around by bus and Tube. Often to be found playing at his local bowls club, he's always the first up to dance at family gatherings.

For the whole family, the scooter symbolises wonderful memories; those days out in Greenwich with pie and mash or sausage and chips, going for donkey rides in Greenwich Park. It makes Belinda cry to think about fixing it, she admits. Everyone would love to see Harry riding the scooter once again. What they would really like is for the scooter to be repaired as close to the original as possible, so it looks brand-new,

The good news, says Dom, is that the tyres are OK. Everything else needs work. The first thing Dom does is dismantle the whole thing. As he removes the surface rust with wire wool, fragments of the original bright red colour of the scooter are visible. Dom sends over the framework to the sandblasters to make everything smooth again.

Meanwhile, he fashions a new mudguard bracket for the back wheel, using the good one as a template, and welds that into place. The wooden elements of the scooter need cleaning and conditioning to make them look like new again too. Dom stabilises the cracks and splits with glue.

When the scooter framework returns from the sandblasters, he repaints it in the original vivid colours of bright yellow and red. After researching the model on the internet, Dom reinstates a very thin pinstripe of black paint on the wooden footboard. He also finds the original maker's mark, Tri-Ang, and using a graphics package, re-creates it in a transfer.

Then the final flourishes are added – new grips for the handles and a set of splendid tassels – and it's time for Harry to take it for a celebratory spin.

Dad's Tractor

A vintage toy pedal tractor from a farming family

A treasured childhood toy belonging to a family with generations of farming heritage is in need of repair, so it can serve as a reminder of a beloved late husband and father who died in a tragic accident at work.

Rob Godber, a third-generation farmer, lost his life at just fifty-two when he fell through a barn roof he was repairing. This toy tractor belonged to Rob as a child and stands sentinel by his grave at home in Derbyshire.

His son Will, at the age of eleven, has an encyclopaedic knowledge of tractors thanks to his dad. He has come to the Barn with his mum, Vanessa, Rob's widow and his two older sisters, Jess and Ellie, in their twenties. Rob's daughters also have very fond memories of growing up next door to the farm, with their dad encouraging them to help with feeding and looking after the livestock – and getting muddy.

They would all love the tractor to be in good condition again, as a tribute to Rob, who loved tractors so much he went to his wedding, and his own funeral, on one. However, his vintage Triang toy pedal tractor, dating back to the 1960s, has taken almost sixty years of knocks and bumps and is in a bad way.

Metalwork expert Dom Chinea steps forward and admits that it's going to be a big job. At first glance the tractor looks complete, but the whole chassis is riddled with rust, the seat is wonky, the steering bent, the pedals eroded and the wheels are completely worn through.

He starts by taking everything to bits. The steel parts are sent away to be blasted to remove all the rust and give Dom clean metal to work with and weld back together. When Dom gets the tractor back, however, he realises he is going to have to make some new tubular parts so that the chassis is strong enough to form the backbone of the tractor.

Dom uses his new tool, a tube bender, to pull and bend the steel tubing perfectly into place, and once the chassis is strong again, the rest of the tractor's problems are easy to sort.

His final task is to pick up the tractor's original colours. Now the rust has gone, he can spot chips of red and green, so he repaints it. But he doesn't take out all the 'dings and dents' because these will help to keep the memory of Rob bashing about on the farm, having fun, alive for his family, who are so happy to see the tractor transformed.

This tractor toy holds special memories of a third-generation farmer dad.

TRIANG

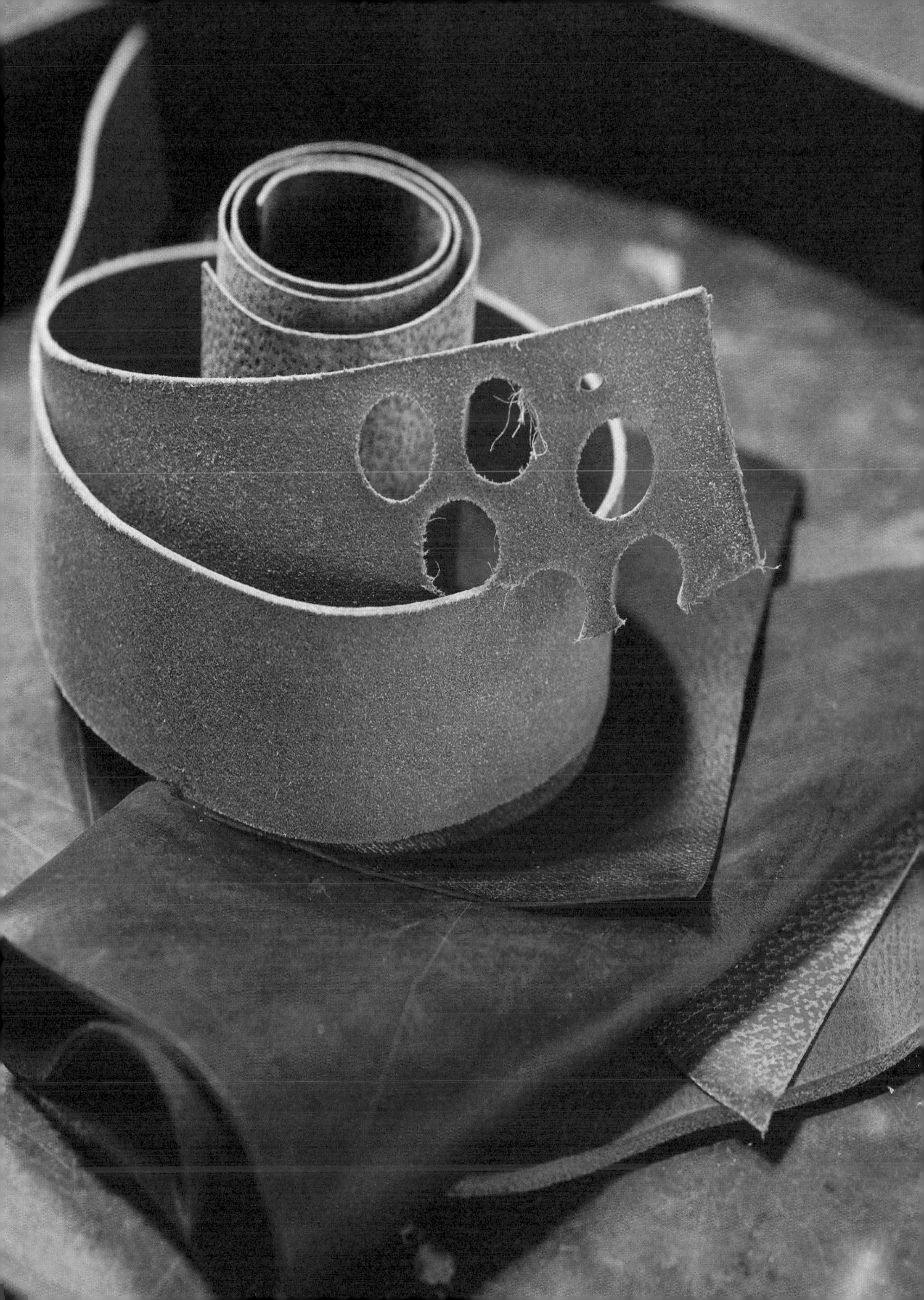

LEATHERWORK

LEATHERWORK

Suzie Fletcher

Leather has many incredible qualities, not least the fact that it is so long-lasting. If properly cared for, items have the potential to survive for decades, if not centuries. The collection of the Museum of Leathercraft in Northampton includes a 3,000-year-old pair of leather underpants that were found in a pyramid, which is an astonishing lifespan.

On the other hand, leather is entirely organic and it is possible to tan it so that it is compostable within a month. At a time when the sustainability of a product is ever more critical, this represents an obvious advantage over plastic, the material most commonly used as a substitute.

The practice of crafting items out of leather dates back to prehistoric times, and was most likely the result of happenstance – an animal blundering into a bog, decomposing, and the vegetation in the bog preserving and tanning the hide. Over time, knowledge must somehow have been gained about how to keep the leather supple by applying a substance such as animal fat. What this leather was used for is a matter for conjecture – clothing and shelter seems probable. From there, it would have been a short step to fashioning all kinds of practical objects.

While harness-making and saddlery are *Repair Shop* expert Suzie Fletcher's particular area of expertise, she can turn her hand to the mending of a diverse range of leather items, from a delicate child's shoe to a battered school satchel. Many of the basic skills involved in leatherwork – such as sewing, cutting, conditioning and filling – are eminently transferable. Learning a craft is not about following a set of instructions like a recipe, it's about acquiring skills that can be employed in a number of different ways. Proof of this can be seen in the way that late nineteenth- and early twentieth-century saddlers, carriage-makers and coach-builders were able to use their skills to fashion and fit out the interiors of motorcars when mechanised transportation arrived.

Leather, a by-product of the food industry, comes from a variety of animal sources and is available in a range of weights and thicknesses. Goatskin and sheepskin are used for making lightweight items, such as gloves, while heavier, harder-wearing articles such as saddles are made from leather derived from bovine hides. After processing, thick hides and skins may be split into two or more horizontal layers, with the top layer being fine grain leather and the bottom layer suede.

If properly produced from start to finish, leather is an expensive material, and so it should be. Cheap leathers are out there, but their provenance is doubtful and problematic. Aside from its production, one of the factors that can compromise leather's quality is its surface finish. Mass-market three-piece leather suites, for example, often have a

surface coating of polymer, a plastic that will never biodegrade. Brightly coloured fashion accessories, such as shoes and bags, are made of split leather to which a surface finish has been applied, and that cannot be maintained in the same manner as naturally tanned leather that has been stained or dyed.

Tanning is the means by which hides are turned into leather and preserved, and nowadays this is usually carried out using chromium salts. Although less commonly employed in modern times, vegetable tanning is a wholly natural process that dates back 5,000 years. It makes use of plant species that are high in tannin, such as oak bark, to create liquors in which the leather is soaked, traditionally for as long as thirty months. Back in the late nineteenth and early twentieth centuries, when there was a vast demand for leather, oak bark was in short supply and tanneries started to use the bark of different trees. These leathers are more likely to develop red rot, which is when the leather disintegrates and turns to dust.

To carry out her work, Suzie has a vast array of tools at her disposal, some of which she has inherited, and she wouldn't want to give any of them up. In the case of edge tools, for example, whose purpose is to trim off the sharp edge of a piece of leather and slightly round it, she has a firm favourite.

When hand-stitching, a key tool is the right awl for the job, and Suzie has a selection in different lengths and widths. In cross-section the blade of an awl is diamond-shaped; as is the case with many tools,

Suzie is the custodian of workbooks that used to belong to a casemaker in 1924.

awls get better and better with use. Once Suzie was reduced to tears, when she overrode her instincts, pushed an awl in at the wrong angle, and broke it. It was one she'd owned for thirty years, ever since she was working at the Royal Mews. Tools need to be used. Suzie intends to pass hers on to The Worshipful Company of Saddlers, so their working life can be extended by a new generation of craftspeople.

Suzie's repair work in the Barn involves making many fine judgements, some as the work progresses, and in particular when the leather behaves in a way she wasn't expecting. While she is quick to point out that she is not a leather conservator, she has learned many of the same skills in her attempts to preserve frail leather treasures. When she's mending items their owners want to carrying on using, rather than simply look at, her approach may change as the work progresses.

When she is crafting something new, it's a question of ensuring that her choices will result in a product that's durable, fit for purpose, and can withstand everyday wear and tear. It all starts with the design and pattern.

Pattern-making is an art in itself. Suzie attributes her ability to think three-dimensionally to her mother, who was a seamstress. To begin with she'll use a material such as canvas or calico to mould a shape – in the case of shoes, for example, she'll mould around a last – an approach similar to making a toile in couture. This will be transferred to card or paper to make a flat pattern for cutting out the leather.

Suzie is a Master English Saddle Maker, who has had apprenticeships with leading saddler Ken Langford and leading harness-maker, William Turner. With all that knowledge, as well as many years of practising her craft, Suzie knows how important it is to pass on such skills to a new generation. A recent member of The Worshipful Company of Saddlers, one of the City livery

MY FAVOURITE TOOL

'Although they aren't particularly special in terms of function, I would be devastated if I lost my granddad's pliers, which I treasure as a tangible reminder of him. He was a clockmaker, and both my brother Steve and I have tools that once belonged to him.

One of my other go-to tools is my head knife. I bought it new about five years ago on the recommendation of another saddler I met at a trade fair. It's very ergonomic and sits perfectly in my hand. The blade is high-quality stainless steel and it holds its edge for a long time. When it does eventually need sharpening, it marries up well with a strop I've owned for thirty-five years. I use the knife for paring, and to cut intricate curved lines as well as straight ones.'

companies, she's involved in its trade and education committee, which works to keep skill centres open and the craft alive. Among these are the leather skill centre at Walsall, and Capel Manor College in Waltham Cross, which offers training in saddlery and footwear.

Before taking up her apprenticeships, Suzie trained at Cordwainers College. Today she is the custodian of workbooks that used to belong to a casemaker who attended evening classes there in 1924 – a casemaker is a leatherworker who produces everything from heavy trunks to hatboxes and purses. Earlier this year she had the opportunity to present this fascinating historical record at an event at the livery hall. Among many intriguing entries are details of a leather hatbox, designed to store a top hat and a bowler hat. Preserving these books is a tangible means of handing down valuable knowledge that might otherwise be lost.

Taking the time to maintain and care for your leather items will keep them in good condition and extend their life.

CARING FOR LEATHER

- Prevention is always better than cure, and taking the time to maintain leather items will keep them in good condition.

- When leather dries out, it loses its natural oils and its fibres shrink. Eventually these will crack and break. Once fibres snap and lose their integrity, you're never going to join them up again, although it may be possible to stabilise them.

- For ordinary maintenance, first clean the leather. Wet a soft cloth soaked in lukewarm water and wring it out. Work over the entire surface, and leave to dry.

- For robust leather items, such as saddles and trunks, you can use saddle soap as a cleanser; it's very mild and contains glycerine, which helps to keep the leather supple.

- The next stage is conditioning. Use a special leather conditioner and apply a thin layer using a soft cloth. Don't use too much, or the leather will stretch out of shape.

- Small holes and gouges can be filled with leather filler, which is a mixture of leather dust and glue.

- Between uses, store leather items in cloth bags out of direct sunlight. Sunlight not only dries the leather, it causes fading and discolouring too.

THE
REPAIR
SHOP

Ukrainian Suitcase

Peter's poignant reminder of his father's Ukrainian roots

Taken from Ukraine by the Nazis in 1942 at the tender age of fifteen, Peter Szuszko's father was made to work forced labour for the German Second World War effort. As he wasn't old enough to join the military, the young man's mid to late teenage years would have been spent toiling in armaments factories, scrapyards and textile mills, often being sent back to a labour camp where he was fed and clothed, but his movements restricted.

Peter, from West Yorkshire, explains that his father, Peter senior, passed away the Christmas before the 2022 Russian invasion of Ukraine. He left behind the leather suitcase in which he had carried everything he owned when he arrived in England after being liberated by the Americans at the end of the war.

The brave traveller docked at the port of Hull with a friend who had heard that the UK was a nice place. Aged twenty at this point, there's a photograph of him on that auspicious day, and leather worker Suzie Fletcher can't believe how young he looks.

Peter says that the current Ukrainian situation resonated; he thought about how his dad had to leave his home country eighty years ago. As Peter carries the suitcase into the Barn, he felt as if he was walking as his dad had.

On arrival in the UK, Peter senior found a job as a weaver, but he was still classed as an alien. Peter shows Jay the documentation his dad kept in his case, as well as a workbook from the day he got to Germany. He says that his dad, who had no formal education, was a nice, funny and intelligent man, a 'grafter' who always worked hard, setting an example to his children.

His precious suitcase however, has become battered over the years. Peter doesn't know how much damage he and his sister did, playing with it as children. The handle, although worn, is especially important to Peter as his dad's own hand held it. There are white paint marks on the case's exterior, probably emulsion, Peter says, because it was kept on top of a wardrobe close to the ceiling.

With a warped lid and a broken front edge, the case is really showing its age, Suzie says. The frame that holds the case together and gives it structure has been bent and that's why the front has become distorted, so Suzie needs to add extra card to give it strength. In order to do this, she is going to have to remove all of the lining.

She had hoped that the lining and first layer of card would come apart really easily, but it hasn't. She dabs the lining with a damp cloth to reactivate the glue so she can pull it away from the card.

With pieces of the case spread everywhere, Suzie dismantles the base. She takes a new piece of card to glue the lining to and this is then glued to the inside of the case's lid to give it structure and prevent everything going out of shape again. An extra strip of leather that matches the case is carefully placed along the edges; she is confident it will look as if it was always like that.

Peter senior used this leather suitcase to carry his belongings to England after the end of the Second World War.

ADDING A NEW HOLE TO A BELT OR STRAP

You can make an extra hole in a belt or the strap of a bag using a leather punch, an inexpensive tool that is widely available online. These have rotary heads that enable you to match the diameter of the punch to the existing holes.

Measure the distance between the existing holes and mark where the new hole should go, taking care to also position it centrally on the width of the belt or strap. Then grip the punch and squeeze to cut a new hole.

Now Suzie can turn her attention to the handle. She puts a new pigskin leather piece on the underside to hold the layers together. Although the hole she makes goes through all the thicknesses of leather, she can't get the needle in all the way, so she employs a technique she was taught when making harnesses; using fishing line to pull through the thread. After trimming the leather, Suzie sands down the edges and adds a stain and polish to buff it all up beautifully.

When Peter returns to the Barn, he is full of trepidation. He says that his dad probably clung onto his suitcase for grim death because it had carried everything he owned.

Suzie points out that someone has attempted a previous repair. If this looks like it has involved glue, Peter laughs, it will definitely have been a job

Suzie strengthens the case with a new lining and touches up the aged leather to restore this suitcase which represents Peter Senior's courageous journey.

undertaken by his dad, who thought glue could repair anything.

But he turns serious when he sees the splendid result of Suzie's painstaking efforts. The restoration of his precious suitcase has amplified the connection between Peter and his dad.

Peter says that every time he picks it up now, he will know that his dad's hand held the case too. The next time he goes away for the weekend, he will use it, and think of his father, and the courageous journey he took as a young man all those years ago.

COLLABORATION

Thatching Tools

Precious memories of a dedicated dad

Beth Brockett is delighted to note that the Barn has its own thatched roof as she explains why she has brought along handmade tools that belonged to her late father, Peter Brockett, who died of a brain tumour in 1996 when she was just sixteen years old.

Peter was a Master Thatcher, working all over the UK and abroad, and recognised as one of the leading experts in this most rural of crafts, generally making and repairing roofs made from straw or reeds.

Beth, who lives in Chester, is an environmental and social scientist. Her interest in sustainability and the environment grew out of her dad's interest in supporting properly-managed reed beds, which are essential for producing quality thatching materials.

As Beth explains, the well-worn leather belt and the flat-headed, paddle-shaped wooden tools, known as 'leggetts', were used by her dad in his work. Peter wore the leather belt to keep his tools close by when he was on a roof.

But sadly, because they have been in storage for years, the belt and tools are now beginning to show signs of age. The leather of the belt has become very brittle and dirty, with areas starting to flake off. In parts, it looks as if it has been eaten or water-

damaged, dotted with white patches that could be mould.

Wood expert Will Kirk and master saddler Suzie Fletcher will take charge of the repair. Their plan is not to restore the tools to as-new condition – because they will never be used to thatch again – but to take a gentle approach to preserve as much of the originals as possible, so Beth and her family can put them on display.

The leggetts are handmade. Peter used them to dress up coatwork (the coat of thatch on a roof) on combed wheat or water reed. He was just seventeen when he got the horseshoe leggett; it was the first such tool he owned. He may have made it himself, Beth is not sure, but it was definitely made to his specifications, using recycled old horseshoe nails, flattened, then hammered into the paddle. But it has cracked, and the metal elements are rusted; Beth is worried it won't survive much longer.

The flat leggett with copper rings he definitely designed and made himself, Beth says, using basic 15mm copper piping. The wood has now badly discoloured from heat or water damage. Most of the small metal circles are missing, the remaining ones badly rusted.

These tools are so personal to her dad, she can't find exact versions in his old 'thatching bible'. They are also an extension of Peter. His profession was very much tied up with his own identity. Peter started his thatching apprenticeship aged seventeen and became a Master Thatcher by his early twenties. Beth tells us he lived and breathed thatching, working with water reed, long straw and combed wheat reed. He even experimented with heather thatching.

These thatching tools hold memories of a Master Thatcher dad and are one of few possessions left to remember him.

He was very well-respected, recognised by English Heritage for his skills and even worked on the roof of the reconstructed Globe Theatre in Southwark, London. Passionate about natural materials, Peter travelled the world working with organisations to learn about sustainability, and training young thatchers in the UK, America and Ireland.

These tools are so special to Beth because her dad left little else behind. After the trauma of losing someone so close at that age, she struggles even to remember him, as she had blocked the experience from her mind. When she discovered his old thatching tools, memories finally flooded back. Beth remembers these were always either in his hand or the boot of the car. He wore the leather belt on every job, his leggetts in hand. He was a great dad, she says, with a strong work ethic – raising four children and building the family house – always having a cup of tea and cracking jokes.

As the twenty-fifth anniversary of Peter's passing has gone by, she would like the tools to be protected forever. Peter now has four grandchildren. Beth feels that the tools would be a good way to talk about their grandad with her nephews and nieces.

Suzie takes on the belt, leaving Will to tackle the leggetts. He takes the same approach to both tools; cleansing and nourishing the wooden paddles, and using a pendant drill with a little wire brush on top, cleaning all the metal parts, really getting into the nooks and crannies to remove every last scrap of rust. This is a first for Will, as he has never worked on thatching tools before. But he really enjoys scuffing the metal and bringing back the shine.

Suzie loves thatching. She had friends who were thatchers and looked after her thatcher friend's leather knee pads, a vital protective bit of gear, when she worked in the village saddler shop. She appreciates how much of a craft thatching is and what hard work it is.

She is really concerned that the leather belt has become so hard it's like cement and could snap into pieces. Suzie finds a moth larvae, busy eating away at a large area of the surface of the leather on one of the holsters. She fills the holes where the larvae has nibbled it. To do this she mixes up her special leather filler made from adhesive, leather fibres and little offcuts, which fills the holes a treat.

It's important to clean and remove any of the leather surface that wants to come off, including any flakes and mould, and treat the surface with a lubricating gel called Cellugel to stabilise it. The belt is left to dry and absorb the treatment for a couple of hours.

Next, a very thorough treatment with saddle soap brings moisture back to the leather and rejuvenates its texture. Once the belt is cleaned, moisturised and supple once more, Suzie creates an arrangement of weights and boards so she can place the belt completely flat, allowing it to dry out straight.

When the belt has dried, she applies polish and buffs up the leather to a lovely sheen. Both tools and belt are now fit for Beth to display in pride of place.

Suzie restores the well-worn leather belt while Will works on the flat-headed, paddle-shaped wooden tools, known as leggetts.

THATCHING

Nothing could be more evocative of a rural idyll than a thatched cottage nestling in the heart of the countryside, or tucked away in a picturesque village. A building style that is often associated with the British Isles, covering roofs with straw, reeds, turf and the like, was once widespread in many regions around the world, from Japan to North America. The craft dates back thousands of years, and its quaintness is entirely our nostalgic perspective; like many heritage crafts, thatching arose out of the need to make a resourceful, practical and economical use of materials that were readily to hand.

Thatching – techniques, tools and materials – has changed very little over the centuries. Once the most common and cheapest form of roofing, it is estimated that about 60,000 buildings in the UK are thatched at the present time, with up to 75 per cent officially listed as of special historic interest. Nearly a quarter of the total are in Devon. Yet the craft of thatching has dwindled dramatically over the last century, with only around 800 practitioners currently engaged in the trade. With thatched roofs needing replacement every thirty to fifty years, this shortage of skilled labour poses a problem for the survival of these buildings in their original form, structures that, apart from cottages, range from farmhouses and manor houses to dovecotes and other farm buildings, including the Repair Shop Barn.

Like many vernacular building techniques, thatching began as a way of making use of those materials that were available locally, which accounts for some of the regional variations that persist today. Thatched roofs in East Anglia, for example, tend to be made out of water reed because it grows abundantly in the region, and these are characteristically angular. Elsewhere, curved shapes prevail and the most common material is straw, nowadays derived from triticale wheat, which is stronger than modern genetically modified wheat varieties. Some master thatchers grow their own straw so they have a ready supply to hand, harvesting the crop with old harvesters because modern combines cause damage to the stems.

In essence, a thatched roof is formed of bundles of straw (or a similar material), which are laid and secured over an A-framed timber roof. It may be single- or multi-layer. A steepish pitch is critical, because it promotes water run-off, and depth is too, with most thatched roofs being about 25–35cm (10–14in) from the outer surface to the supporting framework.

Complicating matters are features such as dormer windows and hipped ends. For the main roof area, bundles are laid parallel to each other, while at the top, or ridge, they are laid in opposing directions and folded over. Ridge design ranges from relatively plain to quite intricate and decorative. Sometimes thatch is covered with a metal mesh to stop birds and vermin from nesting in it. Because the ridge is the summit of the roof and takes the full brunt of the weathering, it often needs to be replaced every decade or so.

A common misconception about thatched roofs is that they pose a fire risk. Insurance records do not bear this out, though, and provided chimneys are kept swept and in good repair, thatched roofs are no more flammable than those made of other materials. However, if a thatched roof does catch fire, from a stray spark, for example, it can be harder to put out because the deep layers of straw tend to smoulder.

More damaging to thatch are overhanging trees that produce sticky sugars and resins, which foster the growth of mould and rot the straw. Thatching also undoubtedly lasts longer in drier regions of the country and in areas that are not highly exposed. For this reason, sod or turf roofs are more common than thatched ones in the Scottish Highlands.

The work of a silversmith or a woodworker has the potential to last for hundreds of years. Not so the thatcher's; the product of their expertise and skills has a much shorter lifespan. Without viable apprenticeships to pass on the knowledge and keep the craft alive, thatching runs the risk of dying out completely.

Grip

Keepsake returning to the Caribbean

> **'Once I have everything prepared, I'd have to say hand-stitching is what puts me in a quiet, meditative state. It's a beautiful place to be.'**
>
> **Suzie Fletcher**

Keithly Brandy arrives with a suitcase that belongs to his mother, Locita. He wants leatherwork specialist Suzie Fletcher to repair the 'grip', as he calls it, for one final voyage, back to the island of Nevis in the Caribbean, the place Locita left nearly seventy years ago as she headed for England as part of the Windrush generation.

Why does he use the word 'grip'? It's called this because you're gripping it when you carry it, Keithly explains. Locita, who was born in 1935, had packed her grip with the bare essentials when she boarded the ship in 1956, en route to Southampton and then Manchester.

She travelled with her 'beau', the man who would become Keithly's dad, Osbourne Brandy. The journey was unpleasant; the seas were rough, and Locita and Osbourne travelled in the belly of the ship, sustained by 'journey cakes' – made from flour, water and butter mixed together in a dough and fried.

It's fair to say that the grip now shows every one of the thousands of difficult miles it has travelled to reach the Barn today.

When she arrived in England, Locita wanted to go into teaching, but she encountered barriers against those of the Windrush generation. All that was available to her were cleaning jobs at first, but eventually she became a social worker and then a local councillor. Keithly remembers that his mum was always helping other people. She was recognised for her community efforts by Tony Blair when he was Prime Minister.

Suzie feels that the grip was damaged all those years ago during the ocean journey. It's made from card, covered with a fabric that was finished to look

Keithly Brandy enlists the help of leatherwork expert Suzie Fletcher to restore his mother's suitcase for one final voyage back to the Caribbean.

like leather. Possibly, the grip slid across the floor during heavy weather and was smashed, because the stitching is broken, especially on the corners.

First, Suzie has to strengthen the card construction, and to do this she is going to have to take drastic measures, removing the original lining paper of the grip in order to repair the outer layer. She's hoping to remove all the lining paper in one complete piece so it can be re-used when a new structure is in place.

It turns into a tough challenge to separate the lining paper from the grip itself. Suzie first tries to reactivate the glue using dry heat, admitting that this process is one where you don't know what will happen until you try.

When dry heat doesn't work, and neither does steam, Suzie formulates an alternative plan. She decides to cover up the troublesome lining paper with a new piece of card covered with its own lining, but keep the original underneath. She does this in such a way that Locita and Keithly can still see the original lining if they want to; the new card can be easily released from the grip's reinforced structure.

Next, Suzie turns her attention to the seams, all of which need to be resewn really tightly, because Locita plans to use the grip for her journey to Nevis. Using linen thread covered with beeswax – which helps to protect the fibres as they are pulled through the hole – she starts by undoing a small section and gradually working her way through all the seams. It's a technique she uses when working on saddles. Sometimes she has to turn to pliers to pull the thread as tightly as possible. Once she has everything resewn and with the structure solid again, she can focus on the metal fittings.

Suzie enlists brother Steve Fletcher to give an antiqued finish to the new lock that she needs to add. He says using a blow torch to darken the lacquer should do the trick, slightly discolouring the shiny finish.

All the rejuvenated bits of the grip slot and slide together well and are glued together for strength and longevity. Suzie's final task is to buff up the exterior, for which she selects an acrylic that she uses on leather to help disguise the damage here.

Suzie thinks it is really delightful that Locita wants to use the case again, because it carries such meaning for her. From setting off with her belongings as a young woman in her twenties, not knowing where she was going or what was going to happen to her, to being able to take such a poignant return journey – from Nevis to England and back to Nevis again – is a lovely idea, she says.

Suzie strengthens the lining and restitches the seams of the suitcase so it's in tip top condition for its poignant final journey back to Nevis in the Caribbean.

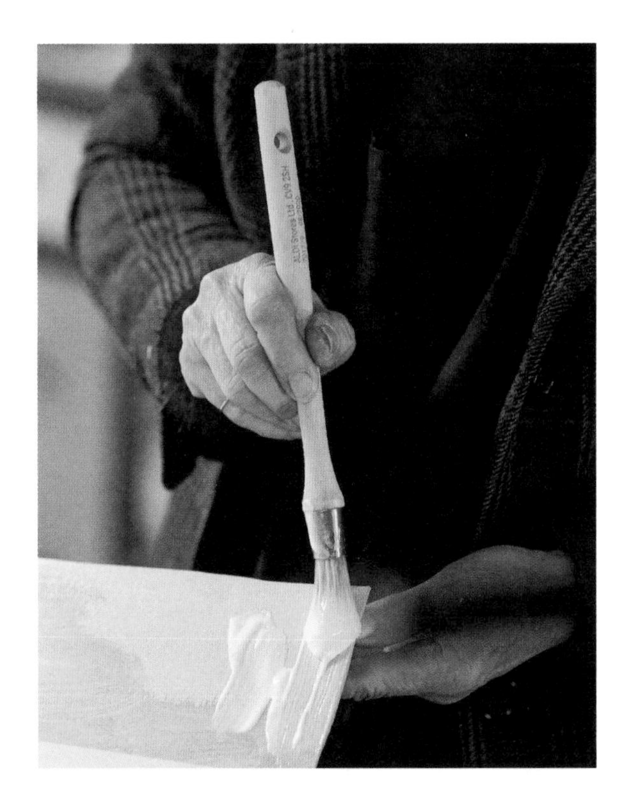

CLOCKMAKING

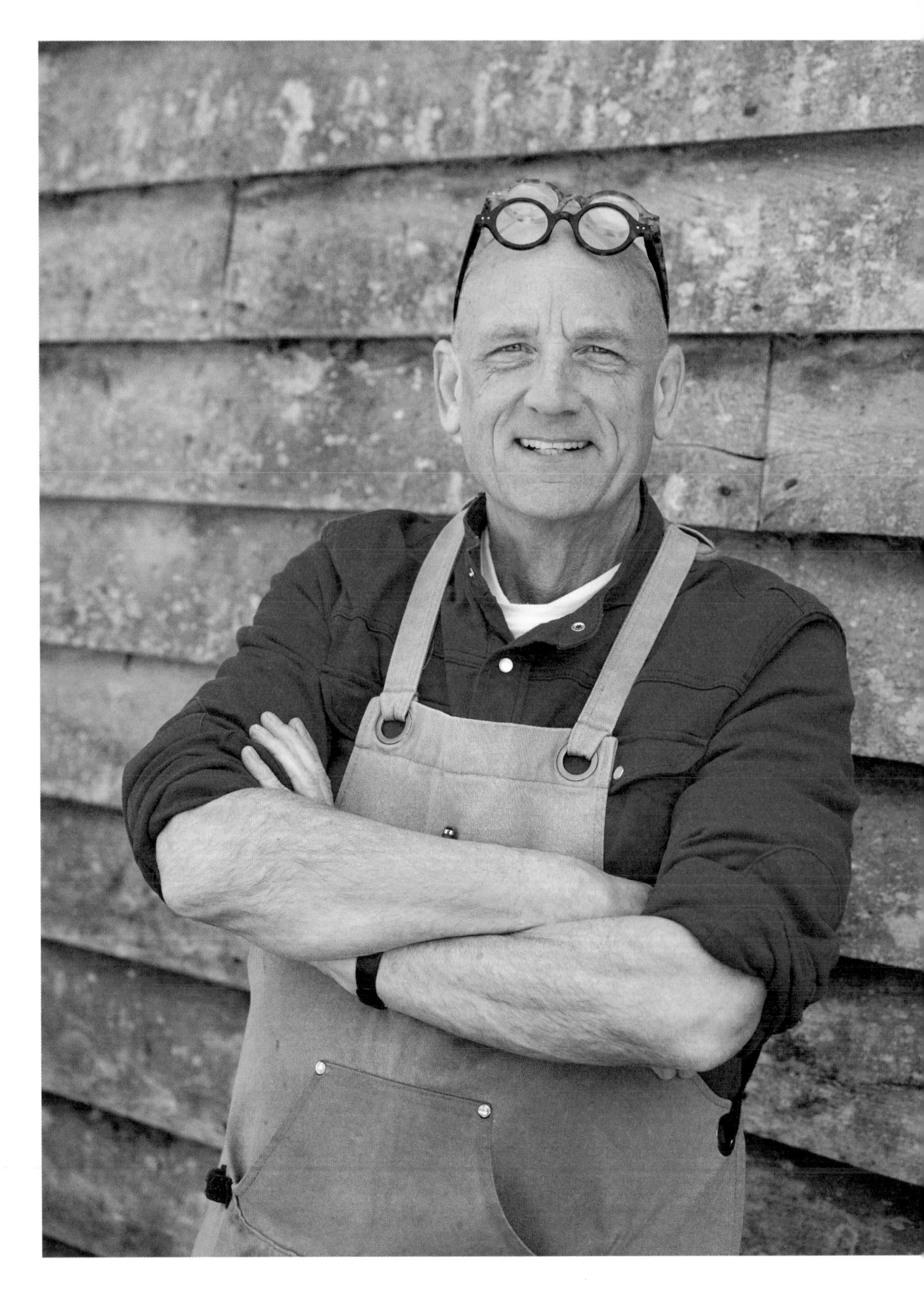

CLOCKMAKING

Steve Fletcher

A ticking clock, marking off the minutes, is the heart of the home, almost a living pulse that animates daily life. We constantly check them to schedule our routines, our comings and goings, and those that ring or chime add a celebratory musical note to the passing of the hours.

It's small wonder that people treasure these objects, particularly if they are heirlooms that have been handed down the generations, or presented to mark a special occasion – like the brass carriage clock traditionally awarded to employees on retirement.

Watches exert a similar tug on the heartstrings, especially those that predate the disposable digital age. While the finest models, which are produced by iconic brands such as Rolex, Cartier, Patek Philippe and Breitling that are status symbols and command high prices, sometimes it's a family association that holds the most value for an individual. Once worn next to the skin, they are intimate keepsakes and reminders of the past.

There's over a century's worth of clockmaking and watch-repairing in Steve Fletcher's family, whose father and grandfather before him were engaged in the trade, and there are many more years to come. Steve's twenty-two-year-old son Fred is following in his footsteps, and his daughter, Milly, manages his business in Witney, Oxfordshire. Crafts die when there's no one left who wants to take up your skills; in the Fletcher family, clockmaking, or horology, is in the safest of hands. In many ways, Steve's family business bucks the trend. Altogether he employs twelve people at his workshop, most of them young, and occasionally he has to stop taking in new work so they can all catch up with the backlog. This is at a time when old-school watch and clock repairers are increasingly shutting up shop, and when he's constantly being offered equipment and spares from businesses that are closing down. Ironically, the problem is time. It can take ten years for an apprentice to get fully up to speed in the craft, time that many repairers just can't afford while they're busy trying to make ends meet. Steve saw first-hand when he was growing up that it can be hard to make a living repairing watches, as his father did, which initially put him off going into the same trade.

Steve generally takes on people who have already been to college and learned the basics. The British School of Watchmaking runs WOSTEP courses in Manchester, which are hands-on and practical courses to train professional watch repairers. West Dean College of Arts and Conservation, located in Sussex near the *Repair Shop* Barn, also has an excellent horology department and Birmingham City University offers a BA course in horology, which covers theory as well as practice. Many well-known watch manufacturers offer training, but this can be less than rewarding since it often involves slotting in modular replacements, which is not the same as the fine skills that are required to repair a mechanical timepiece. Meanwhile, watchmaking is on the Heritage Craft Association's Red List of endangered crafts. Steve regularly gives talks to promote his craft, because what he most enjoys is passing on his own tricks of the trade.

The history of timekeeping devices dates back millennia to 'shadow clocks' or sun dials. Stonehenge is a Neolithic timekeeping monument, the sarsen stones there are precisely aligned to follow the sun's movements. Much later it was Christiaan Huygens (1629–95) who first conceived of using a pendulum as a clock's controller. Many technical refinements have since followed.

Clocks and watches differ in some fundamental ways, scale being the most obvious, but both are mechanisms powered by stored energy. These days this is most often supplied by batteries or electricity, but older timepieces generally fall into two categories: those that are spring-powered, and those that are driven by weights, such as wall clocks and longcase, or grandfather, clocks.

While most people are familiar enough with a clock's outward appearance – the metal, wooden or glass case that encloses the movement or working parts, the face, and the dial and hands that indicate the seconds, minutes and hours – what's inside is often less well understood. The beating heart of a timepiece is the movement, an arrangement of whirring and clicking wheels or gears that transfers the mechanical energy from winding through to the escapement or toothed lever, which gradually releases it, and which is itself regulated by a controller – a balance or pendulum – to keep the time in measured intervals.

In spring-powered clocks, mechanical energy is transferred from a main spring, a coil of tempered metal, to the main train of wheels and pinions. The last gear, or escape wheel, transmits power to the controller, which limits how much is released at any one time. Weight-driven clocks store energy in hanging metal weights suspended from a chain or cord wrapped round a drum. When the clock is wound, the gravitational pull of the weights turns the drum and a ratchet wheel clicks on the main wheel, setting the gears in motion.

Timepieces are complex mechanisms, with small, fiddly components that require precision engineering. Steve's trademark glasses, which he tends to wear

There's over a century worth of clockmaking in Steve Fletcher's family; his son and daughter are also following in his footsteps.

in multiple pairs pushed up on top of his head, are readers of different strengths. As and when, he can pull down the one he needs, in order to scrutinise the tiny working parts and figure out how to fix them.

Steve's repairs always start with a careful process of fault-finding, which sometimes means putting the whole problem to one side and sleeping on it until the back of the brain offers up the solution. The main pitfall is rushing, as you may overlook little faults. Taking time always saves you time in the long run.

To fix a clock may entail making broken or missing parts – from wheels to levers, pivots and shafts – using basic metalworking skills such as filing, cutting and working with a lathe. Steve's not averse to adding new components to old, as long as the process is reversible at some future date. Occasionally, he will harvest components from a 'donor' clock to fix a broken or damaged one. That said, some of his most difficult challenges are clocks that have been unsuccessfully mended at a previous date using bits of other mechanisms.

In addition, disassembled parts may need cleaning to remove dust and grime. Nowadays, to avoid using harmful chemicals, Steve has returned to old-fashioned ammonia-based cleaners. Oils have also changed completely over the years. Once, the oils used to grease gears were whale oils and sometimes he still opens old clocks and smells a tell-tale fishiness. Modern mineral oils work much better.

From mending watches in the pristine dust-free conditions of his workshop, overhauling the mantel clocks at Chequers, the Prime Minister's country house, to scaling the church tower to service its clock near his Oxfordshire home – a perilous task his grandfather also carried out – there's nothing Steve doesn't know about timepieces. This expertise all dates back to a boyhood fascination with how things worked, a Meccano mindset that today lends itself to fixing all manner of treasures that turn up in the Barn. The skills are eminently transferable.

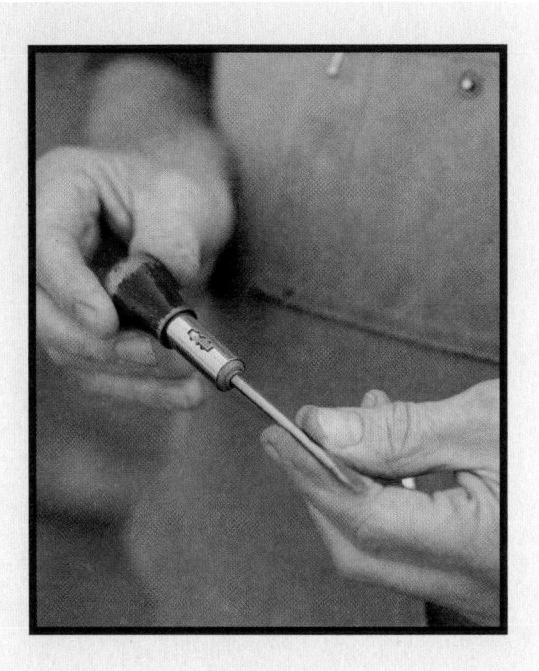

MY FAVOURITE TOOL

'As well as a simple ratchet screwdriver, which belonged to my grandfather, and which just feels great to use, my other favourite tool is a watchmaker's lathe that I inherited from my father. It's a good-quality tool, small and made of steel, which sits on the side of my bench most of the time. I probably use it every day, to turn little components for clocks and watches, among many other things. Other lathes are just not the same. It fits me as comfortably as an old pair of shoes.'

Tavern Clock

A long journey

David Cruickshank drives all the way from Spain to seek help from horologist Steve Fletcher, hoping that a family treasure can be revived.

Steve is so excited because it's a large tavern clock, one of his favourite kind of clocks. He explains that these clocks were placed in taverns when a five shilling a year tax on clocks was brought in by Prime Minister William Pitt in 1797. David knows this story too; people hid away their clocks at home because they didn't want to pay the tax, so they went to the tavern to see what time it was.

This clock has a special place in David's life, because it's always been a part of his family and he remembers it being there since he was very small.

David's father was a farm manager, so the family moved around a lot. Often, they would live in cold, old farmhouses. Although the clock was temperamental, when his mum and dad put it up on the wall it would give him the feeling of 'home'. The historic timepiece became like another member of the family; when he and his sisters could hear it ticking the steady sound was like a heartbeat.

The clock is 240 years old, David believes, and was bought in 1891 by his great-grandfather, Walter Porter, for 20 shillings. It was passed down to David's grandmother, who gave it to his mother before she died. But unfortunately, it has never worked for long. Before David's mother died, he promised her he would have it working reliably again. He says it's a dream, a lifetime's ambition, to fulfil this promise.

Steve identifies that the key challenge is to make it tick again. He's going to concentrate on this and leave the case untouched, because it carries so much history.

When Steve starts to strip down the clock's mechanism he notices quite a lot of the teeth are bent. Now he's got the mechanism apart, he places it all in cleaning fluid. With the wheels, springs and spindle of the tavern clock thoroughly cleaned, Steve straightens the damaged teeth.

The next job is to sort out the repairs to the first wheel. Steve finds that the wheel has been repaired in the past, and there are misshaped teeth which are poorly attached. He heats up the wheel just enough

Tavern clocks are a historical reminder of when tax on clocks was introduced in 1797.

HOW TO LOOK AFTER A CLOCK

■ One of the most popular places to position a clock is pride of place on the mantelpiece. According to Steve, it's also one of the worst, and inadvertently has been responsible for a lot of his trade over the years. An open fire, burning wood or coal, sends minute particles upwards in the rising draught of warm air and this dust is a prime enemy of clockwork. Within three or four years the mechanism will become clogged and cease to function. Instead, position a clock away from the fireplace, out of direct sunlight and on a firm surface where it won't wobble. Keep it wound, but don't overwind it or you'll run the risk of breaking the spring.

■ A clock's movement ticks over millions of times a year. But most people don't send clocks to be serviced until they have broken down, even though they're more than happy to accept the fact that their car needs an annual overhaul to keep it running smoothly. A clock should be oiled every three to four years and thoroughly serviced every eight. Large clocks, such as longcase or grandfather clocks, can go a little longer between services.

■ Don't be tempted to oil a clock yourself. Invariably, you will apply too much or too little, or in all the wrong places. If you use a spray-on oil, you will run the risk of smothering the mechanism and stopping the clock in a week or less.

to melt the solder and the damaged teeth come away.

Repairs on top of repairs are a typical situation with old clocks, Steve says, which is why sometimes the best plan is to take it all off, start again and repair everything properly. He pins two blocks of brass to the good parts of the first wheel and sculpts two new replacement teeth. This results in a complete strong wheel that does the job it was designed for.

He tests that this restored wheel runs into the next wheel nice and smoothly. Steve is delighted when both wheels now work so well together. After giving the clock a second clean, he puts it all back together again. When the back cock (a bracket which holds the rear pivot of the escape wheel) is in place once more, the clock starts to tick perfectly. It's going to be really good for years and years now, Steve says.

All he has to do now is oil everything through, do some adjustments, put the clock's mechanism back into the case and add the pendulum, and it's ready for David to return.

When David arrives, he admits he's been so excited and nervous about the fate of the clock he hasn't been sleeping well. It has been his dream to hear it ticking again, and he was desperate to fulfil his promise to his mother.

Steve asks him if he would like to undo the door at the front of the clock and gently swing the pendulum. As soon as he does, the ticking starts. David says he loves it, absolutely loves it, and can't wait to get the clock home, where he is going to hang it on the wall and look forward to hearing it ticking throughout his house.

Steve sculpts new teeth for the main wheel of the Tavern Clock to help it tick again.

Arctic Thermometer

An intriguing souvenir of a life less ordinary

A broken thermometer from the Arctic Circle that belongs to an adventuring family has recently malfunctioned, possibly due to unseasonably warm summer temperatures.

Former nurse Pat Emmel, 86, and her son, Nick, 59, from Guiseley, near Leeds, West Yorkshire, bring their treasured family heirloom to horologist Steve Fletcher to see if it can be mended.

The thermometer belonged to Pat's late husband, Malcolm, a vicar, who was given the equipment by scientific friends when the Emmel family lived in the Canadian Arctic in the 1960s. Pat and Malcolm moved there as newlyweds and took on a parish. The first year was a challenge Pat tells the Barn; they had no fresh food, lived on tinned produce and their nearest neighbour was over 200 miles away! Pat worked as a nurse, flying out to remote settlements to reach patients, while Malcolm visited his parishioners, baptised babies and officiated at weddings, with the help of the Innuit people and a team of dogs to get around the remote area.

Both Nick and his brother Phillip were born in the Arctic and Nick has fond memories of his family's adventure. His father would delight in reading the thermometer every day and keep a record of the extreme readings.

Malcolm made the wooden casing from a packing case in the Arctic, and Nick is happy for it to retain its original patina.

The thermometer can measure a vast range of temperatures.

Nick doesn't know what's wrong with the treasured piece, which hung in the garage at home, explaining that it recorded the temperature on a very hot day and then it stuck and hasn't moved since. This scientific instrument is a first for Steve. He's intrigued to see how it works but admits he is nervous about dismantling to diagnose the fault. He doesn't want to cause further damage. He suspects it may work using a 'thermocouple' where an internal rod made from two different types of metal creates a voltage between them.

Once he has started to probe, he establishes that the issue is relatively simple. The lever that is connected to the bourdon coil has shifted to one side and jammed and is no longer able to move the indicator hand. Steve loosens a screw to allow him to adjust the lever and tightens it at the correct point so that the lever can move freely once more.

What he now needs to do is test the thermometer alongside a modern electronic one to see if it is reading temperatures correctly.

The electronic thermometer shows the room temperature is 67 degrees Fahrenheit. He sets the Arctic thermometer at the same temperature and puts them both into an oven. What he wants to see is the hand moving up the temperature gauge; it goes slower than the electronic model, but the hand definitely begins to turn. Steve says this is brilliant news.

Next, he needs to test the thermometer at the other extreme; cold. He plunges the sensor end of the thermometer into a bucket of ice and the temperature reading drops rapidly. Steve pronounces the older thermometer to be much better than the modern version.

Adding a bit of oil to the mechanism to ensure it works smoothly, Steve admits he still doesn't know exactly how the thermometer works but his fix to the lever seems to have done the trick.

He's prepared a surprise for Pat and Nick on their return - dry ice. Pat is so pleased to see the thermometer plunging again; it reminds her of the early years of married life. She just wishes Malcolm was there to see it.

This intriguing Arctic Thermometer holds memories of Pat Emmel's early married life in the Arctic in the 1960s.

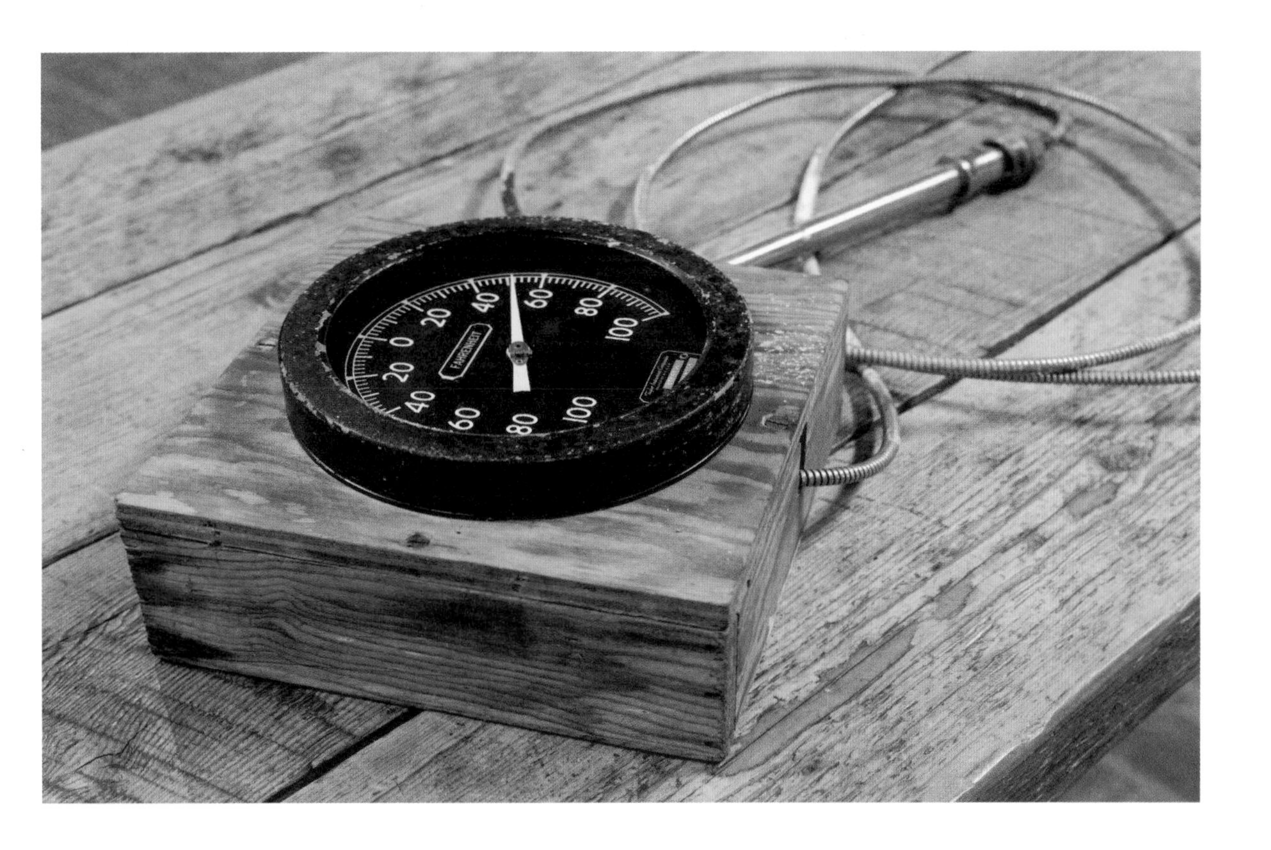

'I regularly give talks to promote clockmaking. What I most enjoy is passing on my own tricks of the trade.'

Steve Fletcher

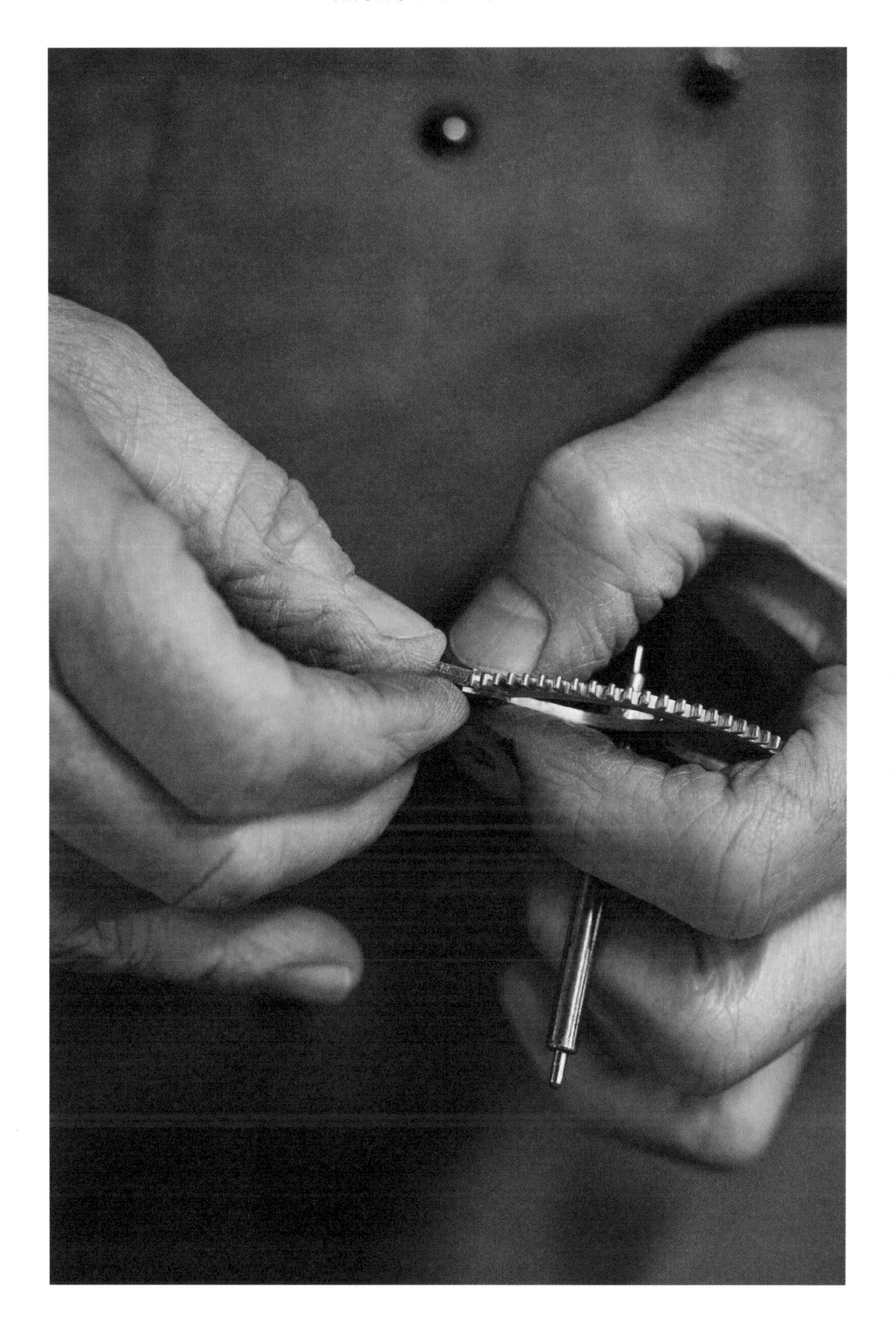

COLLABORATION

Sukeshi's Clock

Travelling in time

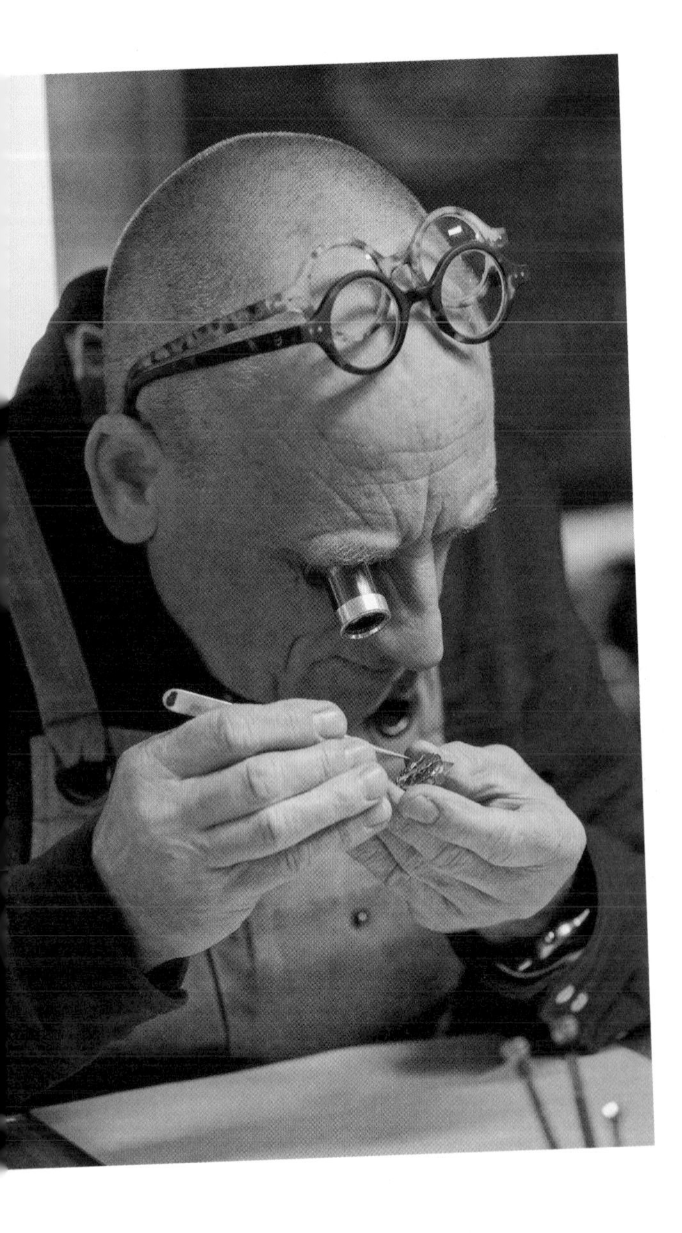

Sukeshi treasured this timepiece ever since her mother gave it to her to aid her independence.

At the tender age of two, Sukeshi Thakkar was given a very special timepiece by her mother, Rama. But a few days after Rama died in 2015, it stopped working. Sukeshi arrives with her carer, Kunda Mangonkar, and horologist Steve Fletcher steps in to see what he can do.

When she was pregnant, Sukeshi's mother was prescribed a morning sickness drug, Thalidomide, which was later found to cause deformation of limbs and other medical conditions in babies. It left Sukeshi disabled, and her mother knew that her child would never be able to wear a standard wristwatch.

But she desperately wanted her daughter to be independent, so she gave her this travel clock, which Sukeshi has treasured as a wristwatch ever since, taking it with her to school, college and university.

Born in Uganda, Sukeshi moved to England with her parents in 1972, because they believed she would have a better life here, and her ever-determined mother joined the fight to gain recognition for Thalidomide children.

Although it hasn't worked since 2015, Sukeshi still keeps the watch on her bedside. She has tried everywhere to get it fixed, without success, and hopes that Steve can help.

He says it's a lovely little thing and he knows how important it is to Sukeshi to get it ticking again. However, the workings contain a multitude of tiny parts that could easily be lost in the Barn, so Steve opts to fix it in his own workshop, within its laboratory-like conditions and he asks restoration expert Kirsten Ramsay to re-paint the case.

When he takes it apart, Steve discovers that some of the teeth on the hour wheel are missing, and

without all of these the watch just will not work. Also, the tiny setting lever spring has a piece broken off. This spring is essential because it connects to the button that sets the time.

He manages to get hold of a new wheel and a new setting lever spring that has the vital piece intact. Very carefully, he pops everything back into place, being careful not to push the delicate parts too hard or he'll end up back at square one.

He's delighted when the watch starts to tick away for the first time in years, and he adds back the dial and hands ready for it to go back into its case. He says he is really looking forward to giving it back to Sukeshi because this watch is such a massive symbol of her mother's love.

Kirsten, meanwhile, has been working to put back the case's enamel detail. When she's buffed the primer that she has put on over the brass finish, she admits she feels quite apprehensive about applying the black enamel. It's such a pretty watch case with lovely brass detailing on it, and she needs to achieve a flat surface with no brush marks. The key, she decides, is to flood the brush with a lot of the enamel paint so it self-levels, a little bit like doing your nails with nail polish. Declaring the end result quite a dramatic transformation, she breathes a sigh of relief.

Steve clips the mechanism back into the case and when Sukeshi returns with Kunda, she is so pleased to see it – and hear it. She says that when the watch is ticking in the night it will be like having her mother's heartbeat back beside her.

Toy Turkey

Farming family's heirloom

Sisters Lynne Timmons from Bishopton, Scotland, and Gayle Rowley, who lives in Bristol, arrive at the Barn with a beloved toy turkey that belonged to their mother, Helen.

They tell Jay and horologist Steve Fletcher that their mum, who grew up on Ballycruttle Farm, a turkey farm in County Down, Northern Ireland, believed that her treasured turkey had been lost forever. However, just before she passed in October 2022, it turned up in a box of old family photos sent by their aunty, Babs Bassett.

The wind-up tin plate turkey, that once walked and flapped its wings, was bought for Helen when she was a child in the late 1950s, but it is thought to have been made in the 1930s by a German company, Ges Geschutzt.

Helen grew up surrounded by animals and livestock, particularly turkeys. Her family would sell turkeys around Christmas time and one of Helen's jobs was to pluck the birds. Lynne and Gail say that their mum loved birds, and she was given this toy by her parents when she was a child and poorly in hospital. She never forgot her very own turkey; it was very special to her because along with a doll, it was one of only two toys she owned. Helen would talk about it all the time when her own children were small, but they had never seen it.

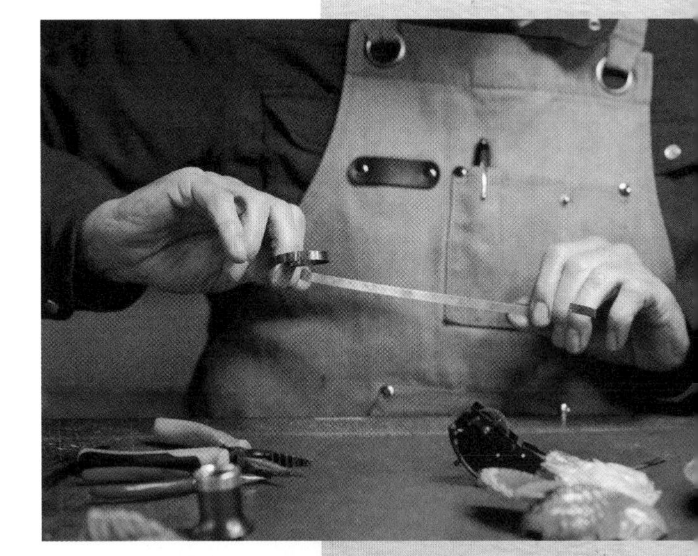

This toy turkey is a reminder of Lynne and Gayle's mother's farming background.

When this mysterious turkey arrived by surprise, Helen was overjoyed to see it again. The family talked about having it fixed, but sadly Helen died before they had the chance to do this.

The sisters have other sentimental items from their mum, but nothing beats the quirkiness of the toy turkey. Lynne says the toy is a link to her mother's upbringing, reminding them all of her farming background. She and Gayle would like to keep this memory alive for generations, by passing it down to Helen's grandchildren, Imogen and Erin. Now her daughters would like to have it mended in her honour.

Seeing it restored and moving around 'with its turkey tail going' will make the family laugh and smile they say. However, due to its age, the turkey is

REPAIRING A MECHANICAL TOY

As long as you're not going to cause permanent damage, there's no reason why you can't have a go at repairing a mechanical toy or similar object. Taking things apart and figuring out how to put them back together again is a good way of learning how things work. A note of caution: many mass-produced mechanical toys have tabs or are seamed and were never meant to be taken apart. If that's the case, it may be better simply to enjoy them as decorative objects.

a little out of shape. Somehow, its feet have become curved and bent so it no longer stands up. It used to walk and, as it did so, its tail would go up and down.

Sadly, the key for the wind-up mechanism is missing and the family are not sure if it even still works. Without the key, they can't tell. They've tried another key, but with no luck.

However, Steve immediately gives them hope. He has a duck toy of a similar type and age, so he's familiar with the workings.

It turns out that the actual body of the toy is in two parts, fixed together with tabs. The body can be opened up by bending these tabs back to reveal the clockwork mechanism. Once Steve knows what's inside, he can work out how to go about fixing it. Cleaning up the clockwork mechanism and dealing with any broken elements is the priority, along with straightening up the turkey's battered feet and fitting a new key.

Steve finds that the teeth of mechanism's cogs are out of alignment and worn, and it is this that is preventing the turkey from working. The blades of the feathers, where they folded over each other, are also rusted and catching on each other.

So Steve carefully takes everything apart. Using wire wool and a rotary dremel tool, he gently scrapes off all the rust clogging up the mechanism and feathers. With a new key and a good clean, the turkey is ready to strut its stuff once more.

'It's weird, but so long as I'm left alone and not interrupted, what I really love is fault-finding. I can get totally absorbed in it for hours – or even days.'

Steve Fletcher

Steve cleans the internal mechanisms and uses a new key to wind up the turkey toy, allowing it to flap its wings once more.

CERAMICS

CERAMICS
Kirsten Ramsay

Ceramic objects of all kinds find their way to *The Repair Shop*, from chipped and shattered vases, plates and figurines, to cracked garden urns and statuary missing limbs or noses. On site is expert conservator Kirsten Ramsay, into whose capable hands these precious family mementoes are delivered for care and sensitive restoration.

Ceramics is an ancient craft dating back millennia, which embraces an enormous breadth of materials, artefacts, decorative styles and finishes, from the simplest coiled pot handmade in red clay to the summit of refinement – delicately glazed and patterned Chinese porcelain. While the earliest ceramic items were clay receptacles for cooking, eating or storing food, baked in the sun or in a fire, the malleability of the material has lent itself to any number of practical and decorative purposes. The oldest clay objects ever found are Paleolithic 'Venus' statuettes, thought to date back to 29,000–25,000 BCE.

The history of pottery and ceramics in general is marked by great technological leaps forward. First, there was the discovery, probably accidental, of firing, then some time between 6,000 and 2,400 BCE the potter's wheel was invented in Mesopotamia. One of the most significant developments was the invention of porcelain, first made in China around 25–220 CE. The method remained a closely guarded secret, known only in the East, until the early eighteenth century when scientists working under the patronage of August II of Saxony cracked the formula and the first Western porcelain was produced at the Meissen factory in Dresden.

Over the centuries, different styles have evolved, often associated with a particular region or technique. Delft tiles, for example, made of tin-glazed earthenware and decorated with characteristic blue and white hand-painted maritime and rural scenes, originated in the Netherlands in the sixteenth century. Wedgwood pottery, whose domestic and decorative ware were embellished with raised white classical motifs, marked the beginning of mass manufacturing in the eighteenth century.
The company established The Potteries, the area around Stoke-on-Trent, as the heart of the burgeoning British ceramics industry. Less than two centuries later, art or studio potters such as Lucie Rie and Bernard Leach helped to revive handmade traditions.

As a craft, pottery makes a broad range of demands on those who practise it, and it requires the mastery of diverse skills. You need dexterity, control and strength to throw and shape pots, an eye for proportion, and an instinct for colour and decorative patterning. Even when such skills are perfected, there are many pitfalls to avoid, some of which are out of your control. Objects can blow up, shatter or crack without warning in the kiln, glazes can be unpredictable and there is often a fine line between disaster and a happy outcome.

What all ceramics have in common is that with the application of intense heat during the firing process they become hard and durable, so much so that earthenware, stoneware and porcelain are the most frequent of all archaeological finds. In fact, much of what we know about prehistoric times comes in the form of ceramic fragments. Another shared characteristic is that ceramics will crack, chip, shatter or break on impact, for example if knocked or dropped on the floor, which is how many beloved family treasures find themselves in *The Repair Shop*.

Kirsten, who has worked on ancient Egyptian artefacts at the British Museum, along with a wide range of Oriental and European ceramics, and whose expertise ranges from terracotta and earthenware to porcelain, enamel, mosaic and glass, is at pains to stress the distinction between craftsmanship and conservation. She locates herself firmly in the latter category, although she admits there is a significant overlap. While her skills embrace some of the same elements of shaping and modelling as that of a potter's, along with a painter's artistry, her focus is on repair and restoration, and as such her skills are much in demand by museums and collectors worldwide.

Kirsten gained her postgraduate diploma in ceramics and related materials conservation at the prestigious West Dean College, Sussex, a stone's throw from *The Repair Shop* Barn. As her private practice grew at her Brighton studio, she started to take on more recent graduates from the same college, sharing the skills she has perfected over the years with a new generation starting out on their careers.

These days, thanks to *The Repair Shop*, the people Kirsten meets are much more clued up about what her work involves. It's a far cry from when she was starting out, when it was only overhearing a chance conversation between Penny Fisher, with whom she later worked at the British Museum, and another student alerted her to the fact that ceramics conservation was a field of study she might like to pursue.

The first stage in any repair is assessment. For Kirsten, an initial thorough cleaning provides her with the opportunity to closely examine an object, work out roughly when it was made, how it was made and what it is made of, along with the extent of the damage and whether it has been previously repaired. Often a piece will be stamped underneath with a manufacturer's mark or brand – Wedgwood, Clarice Cliff or Poole, for example – which cuts down on the guesswork, but sometimes a little more investigation is required. A better understanding of a piece is helpful for the conservation process.

For the actual cleaning process itself, Kirsten often uses a steam cleaner with a narrow nozzle, a specialist bit of kit originally designed for denture-making. Cleaning with steam prevents over-wetting but is still effective at removing some dirt and staining. Pieces are then left to dry thoroughly.

The next stage – assembly – can't be rushed. A common pitfall is hurrying to stick individual pieces together only to discover when you come to the fifteenth or sixteenth that it won't fit, that you've effectively 'locked it out'. You have to proceed very methodically: it's essentially a jigsaw in three dimensions. In preparation, Kirsten will lay out the pieces on a clean sheet of paper to establish how best to put them together, and to figure out what, if anything, is missing. Sometimes she will try out an assembly using tape before committing to the final bond of adhesive.

A further complication is if there has been a previous repair. Nine times out of ten, this will have been carried out using an unsuitable adhesive. Kirsten is always careful to remove every last scrap of the previous glue using solvents and a sharp scalpel so that the broken edge is perfectly clean. If she didn't

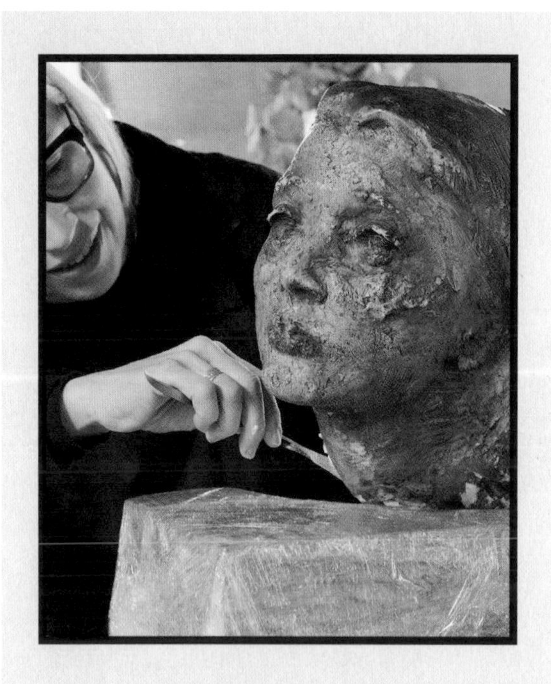

MY FAVOURITE TOOL

'My trusty Tiranti spatula, an Italian modelling tool made of tempered steel, has just the right degree of flex. I bought it from a shop in London when I started studying and I've worked with it ever since. Twenty-seven years later and it's become an extension of the way I think. I use it in countless ways: to apply filler, to mix pigments and to delve into cracks and crevices.'

'Cleaning allows me to really engage with the technical aspect of an object. I'm quite happy to sit and clean for long stretches at a time.'

Kirsten Ramsey

and simply skipped this stage the next join would be very slightly out. Multiply that over many joins and the whole object would end up way out of true. Clear, stretchy magic tape helps to hold the pieces in place while the adhesive cures or 'goes off', in a process that can take days, depending on the surrounding temperature and humidity levels.

Once the object is reassembled, it's time to fill small cracks and crevices and to retouch areas of missing decoration. With experience Kirsten has learned just how much pressure to apply when reassembing objects. Pieces are then filled and consolidated, to prevent vulnerable areas and glaze from crumbling further.

What particularly drew Kirsten to ceramics in the first place was the opportunity to express her great love of colour. As regular viewers of *The Repair Shop* will appreciate, she has an unerring ability to mix and match shades so they blend imperceptibly with original decoration or glazing, creating seamless repairs. This requires not only a great eye, but a painter's meticulous brushwork skills. Her preferred brushes are extremely fine sable brushes (0, 00 and 000 grades) which taper to the fine point necessary for retouching intricate or minute detail. Thicker brushes would make splodges or blobs.

For retouching, Kirsten uses water-based acrylic paint. This might be a specialist clear acrylic medium into which she adds dry ground pigment to arrive at just the right shade and tone. Ground artists' pigments are intense and vibrant and are available from specialist stores; these can also be added to filler. Kirsten also uses artists' acrylics that come ready mixed in tubes.

Naturally, the work of a ceramics conservator demands a great deal of patience. Dexterity, steady hands, an eye for colour and detail are also essential. And since no two jobs are ever the same, so is ingenuity, the ability to problem-solve and to conjure up a whole object out of many puzzle pieces.

Vera's Sculpture

Cherished artistic legacy

Martin Jochman arrives with a gracious plaster sculpture of his mother, Věra, who was born in the former Czechoslovakia and grew up in the shadow of the Second World War.

The sculpture is such a cherished reminder of Věra, and an important symbol of one family's artistic heritage. Touchingly, it was made by an art school friend of Martin's parents, Miloš Axman, who became an important artistic and cultural figure in his country after the Second World War. But over the years the sculpture has become damaged; Martin would be so pleased if ceramic restoration expert Kirsten Ramsay could carry out a repair.

Born in 1922, Věra Jochman (nee Škodová) came from a family of weavers in Upice, in former Czechoslovakia, now Czechia, a small town north of Prague in the Krkonoše Mountains. She was only a teenager when German troops marched into the country. Aged just seventeen, she was sent to work in an ammunitions factory in Berlin.

Recounting the memories his mother shared with him, Martin says Věra actually enjoyed her time in Berlin but eventually planned to escape to Switzerland with two friends. They made it as far south as the border, but there they were caught. The two young men with Věra were shot. She was tried in Nuremberg and sentenced to life in prison.

Věra spent the rest of the Second World War being moved from prison to prison, ending up in Dresden, witnessing the bombing and seeing the sky illuminated red from the fires.

She and her fellow prisoners were liberated by the Russian army as it moved from Berlin towards Czechoslovakia, then she walked all the way from Dresden back to Prague, which was at the time the capital city of Czechoslovakia.

After the war, Věra enrolled at a highly-acclaimed art school in the city of Zlín, where she

This sculpture is an important reminder of Martin Jochman's family's artistic heritage.

studied ceramics and sculpture. There she met Josef Jochmann, a student of painting and illustration. He would later become her husband and Martin's father.

The couple became friends with Miloš, a talented fellow sculpture student. This precious plaster sculpture was one of Miloš' early efforts; he gave it to Věra as a parting gift when they finished their studies.

Martin grew up around the sculpture. It played a formative role in his childhood; he can remember looking at it and playing with it as a little boy. Art is his world, he says, and also part of his family, so the sculpture has huge emotional significance.

Martin continued the artistic family tradition by becoming an acclaimed architect, first visiting the UK in 1968 at the age of seventeen, then studying and settling in Bristol. During a visit to Prague his parents gave him the sculpture.

Věra sadly died in 1999. Although the sculpture had somehow become damaged before that, Martin has taken careful care of it ever since she gave it to him. It is still a beautiful work of art, in his eyes. To see it how it once was would take him back sixty years, bringing back so many memories.

It reminds him of his mother when she was young, and he shares that it looks exactly like his own daughter when she was the same age. The likeness is so striking, that for Martin it represents continuity within his family and their shared heritage. He would love to pass the sculpture on to his daughter one day. This is why he would like it repaired, not just in memory of his mum but to preserve his family legacy.

Sadly, a large area of the neck and hair part has broken away and long since disappeared, along with the wooden pedestal the sculpture was attached to. Some of the plaster is worn and has deteriorated, while other areas of paintwork are damaged.

Kirsten starts by gently cleaning the surface with a soft brush and cotton swabs, working carefully to ensure she doesn't damage the fragile plaster. She then seals the exposed and damaged areas to stabilise the sculpture and prepare it for filling.

Her next challenge is to cast a fixing bracket into the base with which to securely mount it onto its new pedestal, before beginning the next stage of her repair. Kirsten filled the interior with sand, then placed a plastic removable barrier on top. Then using a material called scrim, which is used in architectural plaster work, she built up the missing plaster shape, which hardens and allows her to remove the sand and plastic barrier. The plaster is then consolidated before adding more plaster on top, casting the fixing bracket cast within. The fixing dowel is placed into the sculpture protected with plastic film and plaster is built up around it. Once hardened it can be removed.

Utilising her artist's skills, Kirsten then shapes and sculpts the new plaster so it blends seamlessly with the original. Lastly, she seals the sculpture using PVA and then retouches the bodywork with acrylic paints.

Meanwhile, Kirsten asks woodwork expert Will Kirk to carve a new pedestal, working from photographs of the original to ensure it is a match.

And then, after all their painstaking work, together Will and Kirsten carefully mount the sculpture before returning it to Martin.

Kirsten cleans the sculpture and creates a new bracket to attach it onto its pedestal.

Tower of London Poppy

A military memoriam

Steph has come to the Barn with her step-mum, Alex Briton, to see if her shattered ceramic poppy from the Tower of London art installation that marked the centenary of the First World War can be mended.

Standing on a metal pole, the poppy, part of the installation, Blood Swept Lands and Seas of Red, created in 2014 by artists Paul Cummins and Tom Piper, serves as a poignant reminder of Steph's much-missed late father, Pete, Alex's husband.

The poppy was one of the 888,246 created to commemorate all the lives lost in the war between 1914 and 1918, and which poppies progressively filled the Tower's moat between July and November of 2014.

Pete took Steph, who lives in the South East, to London for the day to see the poppies. Steph recalls that the display was magnificent, and they had a great day out together, enjoying fish and chips and a pint after exploring the installation and the Tower of London.

When the installation ended Pete bought a poppy as a memento of his special day with his daughter, as they were on sale to raise money for service charities. The poppy is so important to Steph and her family because not only is it a special memento of their loved one, but it represents the strong military connection the family share. Pete, Steph and her brother have all been serving soldiers.

Steph joined the Army Cadet Force at the age of thirteen. At sixteen, after leaving school, she went to the Army Foundation College in Harrogate, north Yorkshire, swearing her oath of allegiance to the late

This poppy serves as a special memento of Steph's day out with her late father before he passed away.

Queen at the Tower of London. That's another reason why her connection with the poppy holds such deep significance.

Steph trained with the Royal Artillery and was sent on a tour of Afghanistan in September 2012 for just under seven months. On her return, she struggled to settle back in with her unit. Shortly after visiting the poppies at the Tower of London, and after five years of service, she took the decision to leave the Army and seek a different career whilst she was still young enough to start her life over.

The British Legion charity supported her through the transition from being a soldier to trying to settle back into a civilian life. As the poppy signifies the Army and its veterans, Steph feels strongly about people supporting not only serving personnel but also veterans like herself and her military colleagues.

However, most importantly, the poppy is a cherished reminder of Steph's day out with her dad and is one of only a few possessions she has to remember him by. Sadly, just a month after that memorable day out in London, and after a year of feeling unwell, Pete was diagnosed with amyloidosis. This is a rare condition caused by a build-up of an abnormal protein called amyloid in organs and tissues throughout the body. Symptoms include feeling tired and weak, swelling in the stomach, legs, ankles and feet and skin that bruises easily. There is no cure. The disease progressed quickly; it affected Pete's heart, and he had chemotherapy for eight months, but passed away in a hospice in 2017.

At first the poppy was on display in the family's conservatory. Then they moved it outside. When Pete was having chemotherapy, he was unable to leave the house very often so he spent a lot of time in his beloved garden. He built a stand for the poppy next to the pond, where it looked magnificent, Alex recalls.

However, when Pete went into the hospice, the poppy was left outside, and it stayed there for some time afterwards. Exposed to the elements, its condition began to deteriorate.

Alex says that part of her didn't want to move the poppy from the garden because that's where Pete had placed it. But Steph was worried it would break. When Alex tried to remove it, a petal broke off. Afraid of ruining the poppy with her attempts to mend it, she left it as it was and now feels very guilty for neglecting it for so long.

The ceramic element of the poppy is now broken into between six and eight pieces. Steph thinks all the pieces are there, or if anything is missing it will be a very small piece.

The metal pole that the poppy was attached to needs a good clean and possibly repainting. The paint at the bottom, where the pole was stuck into the ground, is flaking.

Ceramics conservator Kirsten Ramsay says she has repaired quite a few of these poppies in her own studio. Although they are made out of earthenware, the clay has been low-fired, which means they break easily. The red colour comes off too, because the glaze has been lightly baked.

First of all, she assesses the poppy and checks for any missing fragments by laying it out piece by piece. She cleans the surface gently, then seals the exposed broken edges. All the pieces are then bonded together with a quick-drying adhesive, with any missing areas made up with soft filler. Kirsten also adds filler along the break lines. A retouch with acrylic paint provides the final touch for the ceramic element of the poppy, with Kirsten warning that this particular red is notoriously hard to colour match All that's left for metalworker Dom to do is to give the metal pole a clean, and once again this is a poppy to be proud of.

Kirsten cleans the poppy and uses a red paint to match the original colour, to make this special memento as good as new.

Seed Sculpture

Art meets engineering to reunite the pieces of a very special artwork

Maria Quevedo was devastated when she brought back to the UK a marble sculpture carved by her late mother, Betina, an artist who lived in Buenos Aires, Argentina.

Although Maria had carefully packed the precious 'seed' piece – part of a triptych – it was badly damaged on the bumpy plane journey home, causing the marble to come away from its U-shaped wooden base.

Chunks and splinters of wood and chips of marble have broken off, and the two pins that hold the two parts together are no longer functional.

The piece is especially poignant to Maria because the triptych – which includes two further stages that symbolise germination and flight – was the last work her mother created before she died in 2003.

Maria, from London, just wants the sculpture and base to be reunited, and hopes that Kirsten can find a way to do this.

Weighing 25 kilos, the marble is very heavy, Kirsten says, and was attached to the frame by just the two steel little pins. She asks Dom Chinea if he has any thoughts on how it can be fixed securely.

He decides to take the frame apart to get the pins out, with the idea that if he can drill the holes a little deeper to accommodate stronger fixings he will be able to create a more secure fixing. He's confident a solution can be found.

Dom is respectful; he doesn't want to change the look of the sculpture on its stand because he feels that Betina had carefully considered how the two elements would look together. So the first thing he does is unscrew the steel base to remove the pins, which is quite a work-out, he says, but he has the perfect screwdriver.

Meanwhile Kirsten is ready to start bonding the marble fragments back onto the sculpture, choosing a polyester resin because it gives a transparent finish. As the marble has lots of different colours running through it, she explains that it will be possible to 'lose' some of the restoration in the existing detail. When the pieces have all been attached, she can start to fill the missing areas of marble.

This special sculpture is the last piece Maria's mother created before she passed away.

While Kirsten carefully restores the marble, Dom works on fixing the pins which attach the sculpture to its frame.

Dom, meanwhile, has managed to prise the frame apart and has replaced the original pins with longer and wider ones for strength. The fact that the marble is heavy but sits on a relatively minimal wooden frame means some intervention will be required to make sure the sculpture isn't damaged again in the future.

The solution is to attach to the frame a cross-shaped piece of acrylic, with flaps that will bend slightly outwards to support the marble. Dom clamps the acrylic to the frame, using a heat gun to warm it up just enough to bend the material into shape without distorting it. When he's done this, he screws the acrylic support into place on the tallest upright piece of the frame so the sculpture leans into its hold.

As Kirsten fills the marble, she considers what an incredible journey the seed sculpture has taken, from Betina making the piece in Argentina to its arrival in the UK. It will be really lovely for Maria to display this beautiful tactile piece, she says. However, she is worried that the marble itself is not strong and may crumble under pressure when Dom screws into to fit it back onto the stand with the new fixings.

Dom says Kirsten has done such an amazing job of restoring the marble, it's unfortunate he will

WHAT TO DO IF YOU BREAK SOMETHING

'I usually advise people to approach home repairs with extreme caution. Especially if the object is precious or valuable, always seek professional help. Never, ever reach for the superglue. Any repair should be reversible.

Mending ceramic objects is not straightforward because there are so many variables, from the composition of the material to the variety of suitable adhesives, some of which are expensive, particularly those designed for porcelain. The adhesives that conservators use don't yellow and will remain stable for many years if stored in the right conditions.

If you do break something, gather up every single piece, no matter how small. Scout around and make sure you find everything – you don't want to hoover up a crucial fragment at a later date. Then wrap each one individually in tissue or paper towel. This will prevent the edges from abrading further. Store the pieces somewhere safe until you are ready to take the object to be repaired.'

have to drill into it. He chooses a small rotary tool to get into the holes. This will carefully open them out ever so slightly so the new pins he has added to the base slide in snugly and provide a solid fixing. When Kirsten and Dom fit the two pieces back together, the marble fits perfectly into the Perspex cradle. Kirsten is delighted and tells Dom that he's a genius.

Maria returns to the Barn to see how Kirsten and Dom have got on, mentioning that it would have been her mum's 75th birthday just a couple of weeks ago. Although her mother remains a constant presence in her head, she says it would be so nice to have something tangible that she could touch and see to remember her by. As soon as the newly reunited sculpture is revealed Maria just can't stop touching it. She says that by breaking the piece, she put a little bit of herself into her mother's work, but from now on she will make sure it is always on display in her home, in a very safe place.

Ocarina

Music unites the generations

An unusual musical instrument evoking memories of a beloved aunt is in need of repair, because owner Barbara Sykes wants to recreate family musical evenings with her own grandson.

It's an old ocarina, a type of flute made from clay or pottery, that's in an original custom-made leather case. The ocarina was given to Barbara, from Rustington in West Sussex, when she was aged eight or nine, by her favourite aunt, Auntie Joan, her mum's older sister.

Barbara is not sure where her aunt got the ocarina from, but she had a cupboard in her house full of such wonderful objects. It dates from the early 1900s. However, the word ocarina, meaning 'little goose', was first accorded to a musical instrument by Italian brickmaker and inventor Guiseppe Donati, when he invented a submarine-shaped clay flute in the 1850s.

The ocarina wasn't brand new back then; it was already mended and slightly out of tune. The bottom had snapped and was glued back together. She suspects this repair was carried out by Auntie Joan. It's a good repair but one of the notes doesn't play in tune. Also, the gold paint next to each hole has rubbed off so these have almost disappeared. Barbara would like the ocarina to appear better cosmetically, with the paintwork refreshed and notes visible again. If conservator Kirsten can do anything about the crack, that would be great too. But if it's not possible to make it play in tune again, Barbara will understand. After all, it belonged to her dear Auntie Joan and this what it was like when she had it.

Barbara liked to play the ocarina at family gatherings and when she visited Auntie Joan. It was a big musical family and Barbara has many fond memories of evenings together playing their instruments; Barbara's mum Jacqueline and Auntie Joan were two of nine siblings.

She remembers meeting at her uncle, John Phillips', house, with everyone bringing anything that made a noise. Her uncle played a 'Swannee whistle', another type of wind instrument, and there would be guitars, accordions and recorders along with the ocarina.

Sadly, Auntie Joan developed dementia and passed away in 2007. Barbara thinks it would be

This musical instrument is a special piece for a musical family, who hope to play it again.

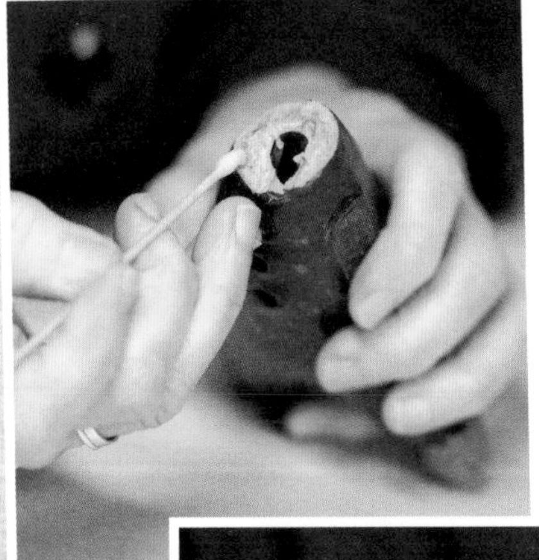

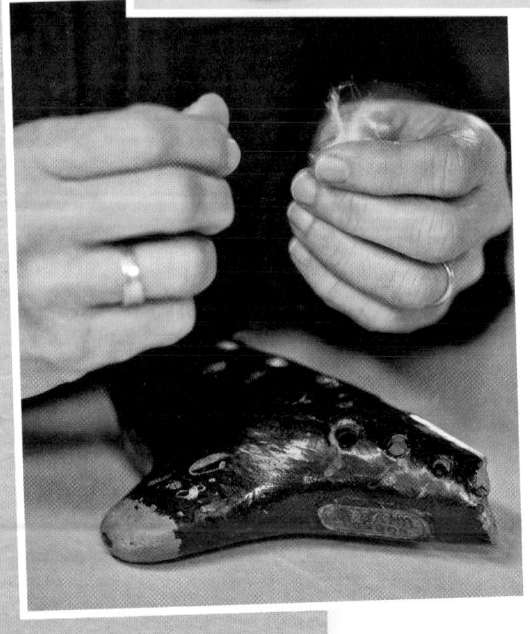

lovely to bring back those family musical evenings they all enjoyed together, especially as her eight-year-old grandson is starting to show an interest in music, learning to play the guitar.

Now that most of the older members of the family have gone – Barbara's mum is the only surviving sibling – the ocarina holds even more significance Holding it is like having Auntie Joan in my hand and it takes me right back to my childhood, says Barbara.

Kirsten knows what an ocarina is and has fond memories of playing them herself as a child. Her mum's best friend lived in a house full of interesting items, including a couple of ocarinas. Kirsten is really excited to start the repair. Amazingly, this is the first time she has ever undertaken a repair on a musical instrument.

First of all, she examines the ocarina to see what it's made of, as this will impact how the repair is carried out. She confirms that it is clay.

The challenge is to undo the old repair without causing damage to the ocarina and its surface decoration. Kirsten can't guarantee that the ocarina will play in tune again, but she is confident that she can get it apart and put it back together to look and sound better.

Taking great care, Kirsten starts to prise apart the repair. She uses acetone in a pipette and a cotton bud to drip the acetone along the crack, to soften and loosen the adhesive. As she wiggles the ocarina to check if there is anything loose she spots a little bit of clay still hanging on – like a wobbly tooth, Kirsten says.

Because the ocarina is made of low fired clay, it's extremely porous, so Kirsten has to seal all the 'break' edges, with diluted adhesive to consolidate them before she can begin to put them together again. When she checks inside the ocarina to see what might be impacting the sound it makes, she couldn't find anything causing the impaired sound. The top

notes are raspy-sounding, not as clear as they should be. If she can carry out the repair successfully, the sound should at least improve.

After leaving the edges to dry, Kirsten uses acrylic filler to mend the crack. She then selects a fine sandpaper and gently sands and smooths the new join so it becomes almost invisible.

With the fix dried thoroughly, she re-paints the filled areas of the ocarina with water-based acrylic paint in natural shades of brown and ochre. When Barbara returns to the Barn, she gives her treasured instrument a try – and plays 'Little Drummer Boy'. All in tune and working again, the ocarina will be the star of the show at the next family musical evening.

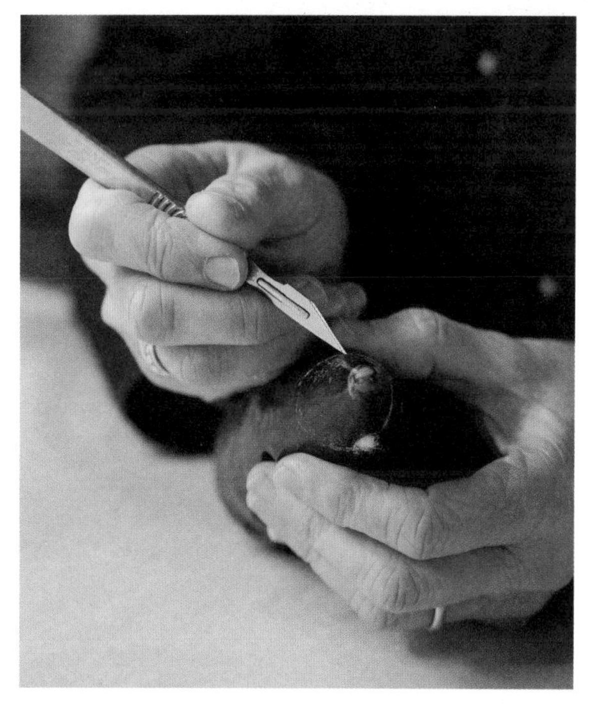

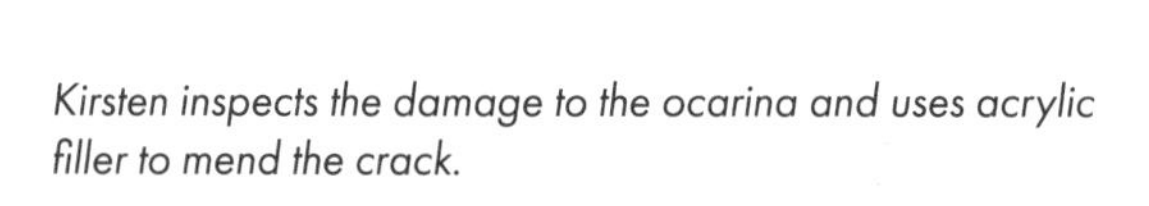

Kirsten inspects the damage to the ocarina and uses acrylic filler to mend the crack.

SILVERSMITHING

SILVERSMITHING
Brenton West

Silver has been treasured for thousands of years. The bright shine and rarity of this precious metal has meant that people have long aspired to collect items made from it. It's these luxurious objects that silversmiths specialise in fashioning; possessions signalling status, wealth and taste, acquired to be shown off and handed down the generations.

Silversmithing, Brenton West is careful to point out, is quite distinct from jewellery-making. Most jewellers couldn't make a silver teapot, he says, and he couldn't set a precious stone. Even so, Brenton's impressive range of skills means that he is not only proficient at working with silver, but with a whole range of other metals, too, from copper and brass to various other alloys, making him an indispensable member of *The Repair Shop* team.

When Brenton originally studied the craft, he counts himself fortunate to have been taught by the very best practitioners. One of his former tutors, Brian Hill, is a trustee and former chair of the Goldsmiths' Craft and Design Council, which promotes both silversmithing and goldsmithing. Unfortunately, very few colleges offer places in silversmithing nowadays, and practical instruction of any kind is seldom on the school curriculum, which means that young people lack the opportunities to explore career paths that aren't purely academic. While Brenton does his best to encourage young students, teaching when he can, he knows *The Repair Shop* is an invaluable showcase, demonstrating where hands-on skills can take you in life. Every year he also donates to the Queen Elizabeth Scholarship Trust (QEST), a charity that awards grants to makers of all kinds and supports excellence in British craftsmanship.

A key stage in any craft is getting to know the material and how it behaves. Silver is very forgiving, malleable and compliant, capable of being formed into various shapes by a whole range of different techniques – by filing, polishing, hammering, cutting, heating and soldering.

When he was a student, Brenton's first task was to learn how to make a seamed bowl, curving a sheet of silver into a cone shape and soldering up the side, then hammering it out into a nicer shape on a dolly, or stake (a curved piece of metal secured in a vice). More difficult was making a raised bowl, hammering a flat disc round and round at an angle to slowly raise up the shape, then using a planishing hammer to smooth the finish. A supreme example of raised silverwork, which is on Brenton's website, is a wine carafe he made from a single disc of silver. With a bulbous bottom and a narrow neck, it stands almost a foot tall.

Silver's malleability also means that it is vulnerable to certain kinds of damage. Drop a silver object from a height, or knock it hard, and it will

dent. Abrade it and scratches will appear. Leave it exposed to air for too long and it will tarnish. When a silver object comes into the Barn for repair, Brenton will polish it to reveal its true condition, before working out a plan as to how to remedy its defects.

For pushing out dents and smoothing finishes, Brenton has an array of hammers at his disposal, with heads of different sizes and shapes, some of which are designed to reach into awkward angles or corners. For cutting, he uses a thin-toothed saw similar to those used by jewellers. To get rid of deep scratches, he uses a file, followed by a succession of wet and dry sandpapers in increasingly fine gauges.

Of all the skills associated with silversmithing, soldering is the most difficult. Silver solder melts at a temperature quite close to the melting point of silver itself. You need all your wits about you and you need to know the signs to look out for. Apply too much heat and you'll melt what you're working on. For this reason, you're supposed to solder silver in the dark, or at least in conditions of low light, so you can spot when the silver starts to glow.

People are often tempted to repair silver items

Brenton enjoys the variety of objects which come into The Repair Shop for restoration and repair.

with lead or soft solder, because it's easy to work and has a relatively low melting point; they're also more likely to have it in their toolkit and to have used it before. But lead solder is extremely corrosive to silver; even a tiny drop will burn a hole right through it. Professional silversmiths don't allow soft solder in their workshops at all. When he was training, Brenton attempted to repair a silver item belonging to a friend's mother and completely destroyed it. When he took it into college the next day to find out what had gone wrong, he was told that the item had lead solder in it.

Knowing that what you're working on actually is silver is one way to avoid many pitfalls in repair. Britannia metal, a white metal that is easy to cast and is used to make budget items, has a low melting point. It can even have marks that look like hallmarks, so it is absolutely imperative that you are familiar with proper silver hallmarks to make a correct identification. Like all crafts, silversmithing means you have to be a problem-solver, adaptable and capable of thinking laterally. Rather than brute strength, you need stamina – prolonged hammering sessions can make your arm ache the next day.

But the rewards are great. Brenton enjoys the variety of objects that come into both *The Repair Shop* and his workshop at home. Recently, he received a letter forwarded by Ricochet, the production company that makes the programme. Simply addressed to 'Brenton West, The Repair Shop', it was from an eighty-seven-year-old woman in Kent, who had been unable to find anyone to fix a silver teapot that had split on the bottom after it had been accidentally dropped. Her husband had bought it second-hand and given it to her when they were married in 1960, and they'd used it five times a day ever since – making approximately 113,150 cups of tea! The repair turned out to be quite straightforward, but their reply was worth its weight in gold: 'Our gratitude to you for its repair is incalculable.'

MY FAVOURITE TOOL

'My favourite tool is my planishing hammer, which I've owned and used for forty-six years now. It's as good as the day I got it, if not better, and it's so familiar to me I could probably pick it out of a box of similar hammers blindfolded. If I lost it and replaced it with the same model, it still wouldn't be the same. It's the feel of it. I bought it when I started training from a company in Clerkenwell which is still trading today.

Planishing hammers, which have highly polished faces, are used for imparting an even finish to metal, so if the face of it is scratched you'll put marks and dents on the metal. From time to time I'll polish mine on a rotary polishing machine to keep the surface perfectly smooth. There's a tiny little nick on one side of the face, but since you only hit with the middle of the hammer, that's not a problem.'

Music Stand

Thank you for the music

A gift that helped to incite a young boy's passion for music and accompanied him throughout his musical career now serves as a reminder of a much-loved and supportive late father.

This foldable steel music stand was given to Will Doyle by his late father, Barry Doyle, but it's now suffering from rust and a broken strap, and the folding mechanism is seizing up.

Barry passed away from a heart attack in his early fifties when Will was only twenty-one. Will was away on tour at the time, but he made it back to the hospital to say his grief-stricken goodbyes.

Barry really was his biggest fan, Will says, and came to every single concert he played. Losing him at a young age was a huge shock and Will regrets that his father didn't live long enough to see his son's career flourish.

He feels so sad that his dad never got to share so many of his musical achievements and special family milestones. He would love to be able to use the music stand again, because it symbolises all the support his dad gave him when he was growing up.

Will grew up on the Isle of Wight and still lives there; his dad was a self-employed carpenter who worked really hard for his family. Along with mum Lola, both Will's parents were super-supportive of their son's passion for music.

When Will was in his early teens his father purchased the stand for him as a surprise gift to support his music. Barry was into antiques, and often went to car boot sales and bric-a-brac markets. He spotted the music stand in an antique shop.

Will's parents did not have a huge amount of money, so he knows how much of a sacrifice buying this music stand must have been. Barry was told when he bought it that this was an original music stand once used by the Royal Marines Band in the 1950s.

This was especially poignant, as at that time Will's ambition was to join the Royal Marines Band Service.

This music stand is a memory of a much-loved father and a musical career.

THE
REPAIR
SHOP

Joining the Royal Marines Band Service never worked out for Will, but his passion for music grew and eventually became his career.

Will went on to study euphonium and trombone at the prestigious Royal Welsh College of Music & Drama. During this time, he became principal euphonium with the National Youth Wind Orchestra, and also joined the Tredegar Band, a world-renowned brass band. His talent took him to perform at every major banding contest. He is now a deputy headteacher and teaches music at a secondary school.

Until the stand became too fragile to travel, Will would take it to concerts and events. Eventually, he could only continue to use his father's thoughtful gift when practising at home.

Will would love to be able to again take the stand to concerts and his music classes in the hope of inspiring a new generation of musicians. Having it restored would also mean he had a constant reminder of the wonderful man that his dad was.

Over the years, however, the metalwork has degraded and there are some areas of rust, which have recently started to spread. One of the two adjusting screws has broken off and the leather strap that holds the stand together for transportation has broken, with only half remaining.

Silversmith Brenton West takes a look, and brings in musical instrument whizz Peter Woods and leather expert Suzie Fletcher to help with the leather strap. Pete produces a new wing bolt for the stand, which is the same kind that's used to adjust drum kits. Pete also confirms that the music stand is official Ministry of Defence issue, of a model that these days is quite rare and valuable. It was quite a find that Will's dad came across in that antique shop all those years ago.

Brenton shares that although his mum was musical, and became secretary of The Bach Choir, in London, and his son is also amazingly musical, their talents have evaded him. But he is good at listening to music, he adds, with a smile.

To make the music stand functional again, Brenton polishes the chrome by hand using a chrome cleaner, and lubricates all the joints. He warns that the stand will not look like new as some of the chrome has come off; but at least it will be clean and not rusty.

He finds that the mechanism has completely jammed. Fixing it is a time-consuming process; Brenton patiently rubs it down with coarse wire wool, using wax to penetrate into the bolts. He gives the chrome a good polish too and it comes up surprisingly well.

To finish the job, Suzie makes a replacement carry strap. She takes width measurements, stains the leather and connects the carry strap back to this well-loved music stand, which now happily unfolds and re-folds with ease.

'Hammering is what puts me in the zone. You get yourself comfortable and then just start tapping away. It's very satisfying. The sound isn't to everyone's taste and it can affect your hearing. I use ear protectors.'

Brenton West

Brenton polishes the music stand and lubricates the joins so that this special object can function again as originally intended.

Army Service Medal

Proud to serve

An Army medal from the Windrush era, representing pride and strength in the face of adversity, needs help from silversmith Brenton West and a flexible hand from teddy bear expert Amanda Middleditch, who will pitch in to replace the frayed and worn ribbon. The medal belongs to Gordon Murrell, from London, who has brought it to the Barn with his daughter, Kelly. He would love to have it back in tip-top condition in time to wear on his eightieth birthday, which is coming up soon.

It's a campaign service medal, awarded to Gordon because he served with the British Army in Germany from 1962 to 1966, following by a nine-month stint in Aden, in the Middle East.

Gordon's life has certainly taken him all over the world. He was born in 1943 in Barbados in the West Indies. Just before his nineteenth birthday, he saw advertisements to join the Army in numerous places; the newspaper, posters, even the famous Lord Kitchener poster from the First World War.

After the Second World War, Britain needed people to join the Army for various aspects of industrial work, Gordon explains. For him, it offered a new experience, an opportunity to see the world. He recalls that the recruitment team impressed him and made him feel that he would fit in.

Sadly, when Gordon arrived at Gatwick, he realised that he would have to acclimatise, and not just to the cold weather. He was stationed in Newcastle, and recalls that in those days some of the locals were not that welcoming.

He and his fellow West Indian soldiers would avoid pubs and clubs in case of trouble, and there were signs that said: 'No Blacks, No Dogs, No Irish.' The racism he encountered was a shock, and not something he had experienced in his own country.

When Gordon was 28, he met Esther, his future wife, who was a nurse in the NHS. They eventually settled down in West London. Life was hard but

This army service medal represents the pride which Gordon felt for serving in the British Army.

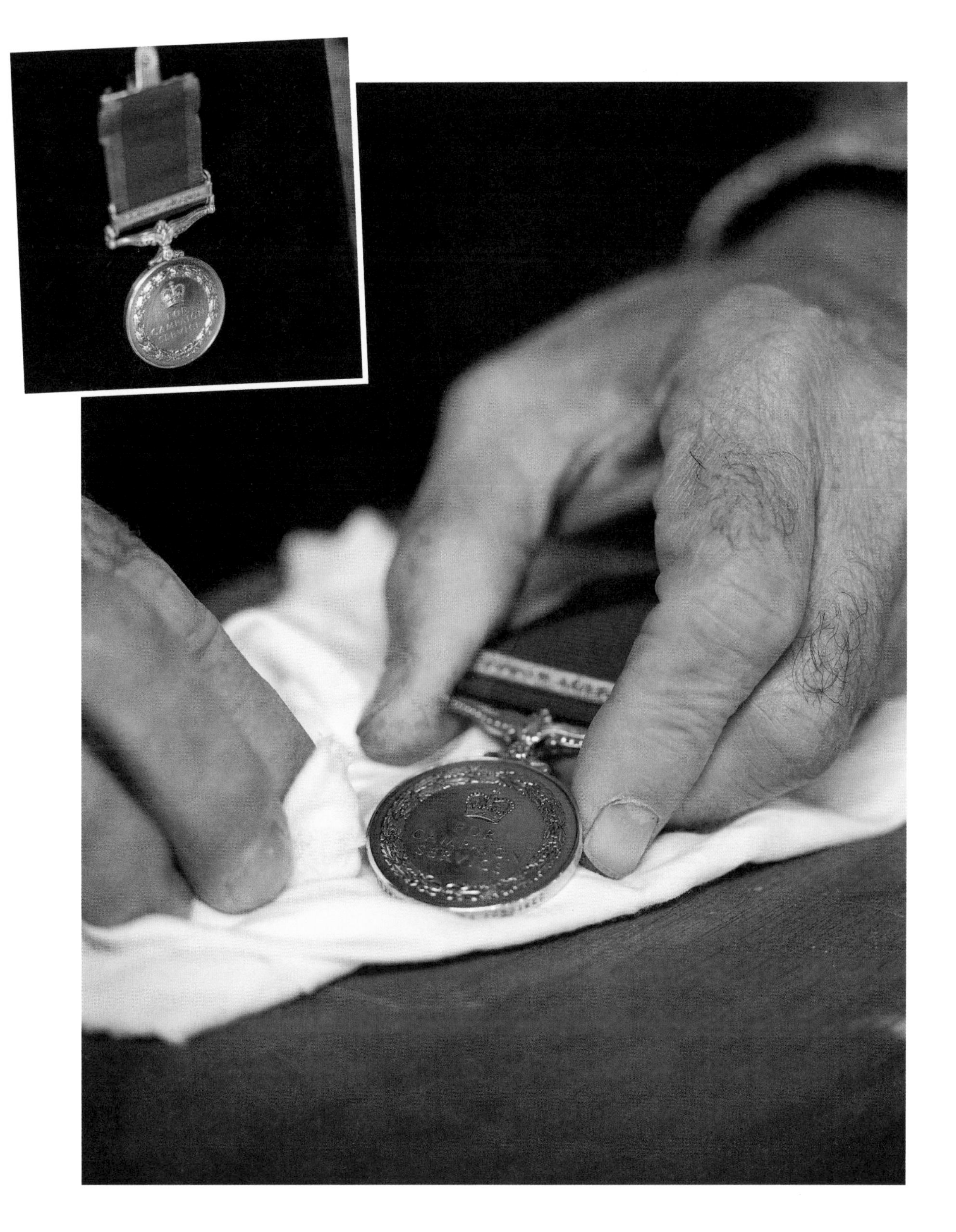

together they stuck it out. There was lots of racism in the Army too, Gordon recalls, but his pride made him not want to return to Barbados. People back home still held Britain in high esteem, and without the knowledge that you could struggle in England, they would think you were a failure if you went back.

This medal represents the pride and sense of achievement that Gordon felt, because it shows that he served in the British Army and has something to show for it. This is made especially poignant as such medals are no longer issued.

Over the years, the medal has suffered from wear and tear. The ribbon is broken, so it's in two parts. Gordon says he would be honoured if Brenton could do something about the dents and scratches to make the medal look as it did originally, and polish up the image of the late Queen Elizabeth II on the front.

Kelly agrees, and says it would be a great honour to see this piece of living history revitalised. It would help her come to terms with the fact that one day her dad may no longer be here, but she will have a lasting piece of him to hang on to. As well as serving as a connection between father and daughter, it would be a remembrance of his sacrifice when coming from Barbados to England too.

Brenton carefully cuts off the original ribbon and hands it over to Amanda. The medal is silver, he discovers, and he thinks it looks as if it has been dropped,which probably caused the damage.

The lip of the medal has dented in, at the spot where Gordon's military number – which he can still recite – is engraved. This is such an intricate job, that

Brenton fixes the dent in the edge of the medal and polishes it to its former glory, while Amanda finds a replacement ribbon.

CARING FOR SILVER

Silver tarnishes with prolonged exposure to air. If you don't intend to display an item, keep it in a cloth bag or airtight container. Silverware should be washed by hand, never in a dishwasher, and stored in felt-lined canteens or drawer dividers. Never use silver cutlery to eat eggs – the sulphur in the eggs will blacken the metal.

Always use silver wadding polish on silver – polish designed for other metals, such as brass, for example, is much too abrasive. Rub gently with a soft cloth, rinse, then buff dry. Be especially gentle if you are polishing silver plate. If you apply too much pressure, you run the risk of polishing right through the plating.

solder can't be used. So Brenton makes a specially designed brass punch – a long piece of brass with a punching tool on the end – and repeatedly taps the metal over and over again, until the rim looks complete again.

Meanwhile, Amanda has sourced a new identical ribbon which she hand-sews back onto the medal. This new ribbon is so crucial because it is the only way that the medal will be able to still be worn by Gordon on his birthday and at Remembrance Day and other ceremonial occasions. Touchingly, the original ribbon is preserved and returned to Gordon for his safe keeping.

Amanda does a lot of research to find an exact replacement ribbon; each colour in the fabric represents the country Gordon is from, the regiment he served in and the campaigns he took part in. She is also careful about making sure the ribbon is the exact specific length; all medals, when worn together, have to line up correctly. Finally satisfied, Amanda sews the ribbon firmly back onto the metal. When Brenton has given it a final polish, it is ready to return to one proud veteran soldier.

Index

Thank You

Thank you to *The Repair Shop* team who worked on this book: Hannah Lamb, Paula Fasht, Emily Senior, Jade Kitson, Helen Page, Sarah Wilson and Finley Thomson.